Wavelets

Algorithms & Applications

Wavelets

Algorithms & Applications

Yves Meyer
CEREMADE
and
Institut Universitaire de France

Translated and Revised by
Robert D. Ryan
Office of Naval Research
European Office

Society for Industrial and Applied Mathematics
Philadelphia   1993

Library of Congress Cataloging-in-Publication Data

Meyer, Yves.
    [Ondelettes. English]
    Wavelets : algorithms and applications / Yves Meyer ; translated
by Robert D. Ryan.
        p. cm.
    "Translation based on lectures originally presented by the author
for the Spanish Institute in Madrid, Spain, February 1991"—CIP
galley.
    ISBN 0-89871-309-9
    1. Wavelets. I. Title.
QA403.3.M4913   1993
515'.2433—dc20   93-15100

10 9 8 7 6 5 4 3

siam. is a registered trademark.

Contents

Translator's Foreword

The question most often asked by those who heard I was translating this book was, "How did you get the job?" Well, I asked for it. I mentioned to Professor Meyer in March 1992 that I had heard that he had written a new book. He said, yes, and that the book was based on notes from lectures he had given at the Spanish Institute in Madrid. He added that the book was being translated into Spanish and that SIAM was interested in publishing an English edition, for which they would need a translator. I volunteered to do the job; what you have is the result.

In addition to translating the text, I have tried to "work through" much of the mathematics to correct typos. I have also added a line of explanation here and there where it seemed appropriate, and sections of text and references have been updated. These revisions are not highlighted in any particular way (i.e., there are no translator's notes except at one reference), but rather incorporated into the text. These changes were made with Professor Meyer's blessing. Of course, there is always the possibility that in the process of updating the manuscript I have introduced other errors; for these I take full responsibility.

The great fun of this project has been the chance to work with Yves Meyer and other members of "team wavelet." Professor Meyer improved both my French and my mathematics, and his enthusiasm and appreciation for my efforts kept things moving. Direct help also came from John Benedetto, Marie Farge, Patrick Flandrin, Stéphane Jaffard, and Hamid Krim. Alex Grossmann gave moral support by assuring me that it was an important project. My sincere thanks to all of these people. The work was done while I was a Liaison Scientist for the Office of Naval Research European Office, where my primary job was to report on mathematics in Europe. It was through this work that I first made contact with the French wavelet community, and I thank the Office of Naval Research for that opportunity. This was essentially a weekend and evening project, and hence a family project. In this context, I thank my son, Michael J. Ryan, who kept house and produced great dinners while I kept the electrons moving.

Robert D. Ryan
November 1992
London, UK

Preface

The "theory of wavelets" stands at the intersection of the frontiers of mathematics, scientific computing, and signal processing. Its goal is to provide a coherent set of concepts, methods, and algorithms that are adapted to a variety of nonstationary signals and that are also suitable for numerical signal processing.

This book results from a series of lectures that Mr. Miguel Artola Gallego, Director of the Spanish Institute, invited me to give on wavelets and their applications. I have tried to fulfill, in the following pages, the objective the Spanish Institute set for me: to present to a scientific audience coming from different disciplines, the prospects that wavelets offer for signal and image processing.

A description of the different algorithms used today under the name "wavelets" (Chapters 2–7) will be followed by an analysis of several applications of these methods: to numerical image processing (Chapter 8), to fractals (Chapter 9), to turbulence (Chapter 10), and to astronomy (Chapter 11). This will take me out of my scientific domain; as a result, the last two chapters are merely resumes of the original articles on which they are based.

I wish to thank the Spanish Institute for its generous hospitality as well as its Director for his warm welcome. Additionally, I note the excellent organization by Mr. Pedro Corpas.

My thanks go also to my Spanish friends and colleagues who took the time to attend these lectures.

CHAPTER 1

Signals and Wavelets

The purpose of this first chapter is to give the reader a fairly clear idea about the scientific content of the following chapters. All of the "themes" that will be developed in this study, using the inevitable mathematical formalism, already appear in this "overture." It is written with a concern for simplicity and clarity, while avoiding as much as possible the use of formulas and symbols.

Signal and image processing always leads to a collection of techniques or procedures. But like all other scientific disciplines, signal and image processing assumes certain preliminary scientific conventions. We have sought, in this first chapter, to describe the intellectual architecture underlying the algorithmic constructions that will be presented in the following chapters.

1.1. What is a signal?

Signal processing has become an essential part of contemporary scientific and technological activity. Signal processing is used in telecommunications (telephone and television), in the transmission and analysis of satellite images, and in medical imaging (echography, tomography, and nuclear magnetic resonance), all of which involve the analysis and interpretation of complex time series. A record of stock price fluctuations is a signal, as is a record of temperature readings that permit the analysis of climatic variations and the study of global warming. This list is by no means exhaustive.

Does there exist a precise definition of a signal that is appropriate for the field of scientific activity called "signal processing"? A needlessly broad definition could include the sequence of letters, spaces, and punctuation marks appearing in Montaigne's *Essays*, but the tools we present do not apply to such a signal. However, the structuralist analysis done by Roland Barthes on literary texts shares some amusing similarities with the multiresolution analysis that we describe in Chapter 4.

The signals we study will always be series of numbers and not series of letters, words, or phrases. These numbers come from measurements, which are typically made using some recording method. The signals ultimately appear as functions of time. This is true for one-dimensional signals. The case of two-dimensional signals will be examined in a moment.

The objectives of signal processing are to analyze accurately, code efficiently, transmit rapidly, and then to reconstruct carefully at the receiver the delicate

*oscillations or fluctuations of this function of time. This is important because
all of the information contained in the signal is effectively present and hidden in
the complicated arabesques appearing in its graphical representation.*

These remarks apply to speech: A speech signal originates as subtle time
variations of air pressure and becomes a curve whose complex graphical characteristics are an "adapted copy" of the voice.

It is equally important to consider two-dimensional signals, which is to say,
images. Here again, image processing is done on the numerical representation of
the image. For a black and white image, the numerical representation is created
by replacing the x and y coordinates of an image point with those of the closest
point on a sufficiently fine grid. The value $f(x, y)$ of the "gray scale" is then
replaced with an average coefficient, which is then assigned to the corresponding
grid point.

*The image thus becomes a large, typically square, matrix. Image processing
is done on this matrix.*

These matrices are enormous, and as soon as one deals with a sequence of
images, the volume of numerical data that must be processed becomes immense.
Is it possible to reduce this volume by considering the "hidden laws" or correla-
tions that exist among the different pieces of numerical information representing
the image? This question leads us naturally to define the goals of the scientific
discipline called "signal processing."

1.2.   The goals of signal and image processing.

Experts in signal processing are called on to describe, for a given class of signals,
algorithms that lead to the construction of microprocessors and that allow certain
operations and tasks to be done automatically. These tasks may be: *analysis
and diagnostics, coding, quantization and compression, transmission or storage,
and synthesis and reconstruction.*

We will use several examples to illustrate the nature of these operations and
the difficulties they present. It will become clear that no "universal algorithm"
is appropriate for the extreme diversity of the situations encountered. Thus, a
large part of this work is devoted to constructing coding or analysis algorithms
that can be adapted to the signals that one processes.

Our first example is the study of climatic variations and global warming.
This example was discussed by Professor Jacques-Louis Lions at the Spanish
Institute in 1990, and the following thoughts were inspired by his talks [4].

In this example, one has fairly precise temperature measurements from different points in the northern hemisphere that were taken over the last two centuries,
and one tries to discover if industrial activity has caused global warming. The
extreme difficulty of the problem arises from the existence of significant natural
temperature fluctuations. Moreover, these fluctuations and the corresponding
climatic changes have always existed, as we learn from paleoclimatology [6].

To specify a *diagnostic*, it is essential to *analyze*, and then to *erase*, these
natural fluctuations (which play the role of noise) in order to have access to the

"artificial" heating of the planet resulting from human activity. The diagnostic often depends on extracting a small number of significant parameters from a signal whose complexity and size are overwhelming. Thus the analysis and the diagnostic rely naturally on *data compression*. If this compression is done inappropriately, it can falsify the diagnostic.

Data compression also occurs in the problem of *transmission*. Indeed, transmission channels have a limited capacity, and it is therefore important to reduce, as much as possible the abundance of raw information so that it fits within the channel's "bit allocation."

One thinks, for example, of the digital telephone (Chapter 3) and the 64 Kbit/sec standard, which limits, without appeal, the quantity of information that can be transmitted in one second.

A more surprising example appears in neurophysiology. The optic nerve's capacity to transmit visual information is clearly *less* than the volume of information collected by all the retinal cells. Thus, there must be "low-level processing" of information before it transits the optic nerve. David Marr has developed a theory that allows us to understand the purpose and performance of this low-level processing. We present this theory in Chapter 8.

We now consider problems posed by *coding* and *quantization*. Different coding algorithms will be presented and studied in this work: subband coding, transform coding, and coding by zero-crossings. In each case, coding involves methods to transform the recorded numerical signal into another representation that is, depending on the nature of the signals studied, more convenient for some task or further processing. Quantization is associated with coding. The "exact" numerical values given by coding are replaced with nearby values that are compatible with the bit allocation dictated by the transmission capacity.

Quantization is an unavoidable step in signal and image processing. Unfortunately, it introduces systematic errors, known as "quantization noise." The coding algorithms that are used (taking into account the nature of the signals) ought to reduce the effects of quantization noise when decoding takes place. One of the advantages of *quadrature mirror filters* is that they "trap" this quantization noise inside well-defined frequency channels. These filters will be studied in Chapter 3.

The problems encountered in *archiving* data (as well as problems of transmission and reconstruction) are illustrated by the FBI's task of storing the American population's fingerprints. Different image-compression algorithms were tested, and a variant of the algorithm described in Chapter 6 gave the best results. This established a standard for fingerprint compression and reconstruction.

The last group of operations consists of *decoding, synthesis*, and *restoration*. Synthesis and decoding are the inverse operations of coding and quantization. The task is to reconstruct an image or audible signal at the receiver from the series of 0's and 1's that have traveled over the transmission channel. One thinks of decoding an encoded message, as in cryptography, and this analogy is correct because one cannot reconstruct an image or signal without knowing the coding algorithm.

Signal restoration is similar to the restoration of old paintings. It amounts to ridding the signal of artifacts and errors (which we call noise), and to enhancing certain aspects of the signal that have undergone attenuation, deterioration, or degradation.

1.3. Stationary signals, transient signals, and adaptive coding algorithms.

We have just defined a set of tasks, or operations, to be performed on signals or images. These tasks form a coherent collection. The purpose of this book is to describe certain coding algorithms that have, during the last few years, been shown to be particularly effective for analyzing signals having a fractal structure or for compression and storage. We will also describe certain "meta-algorithms" that allow one to choose the coding algorithm best suited to a given signal. To better approach this problem of choosing an adaptive algorithm, we briefly classify signals by distinguishing stationary signals, quasi-stationary signals, and transient signals.

A signal is stationary if its properties are statistically invariant over time. A well-known stationary signal is white noise, which, in its sampled form, appears as a series of independent drawings. A stationary signal can exhibit unexpected events, but we know in advance the probabilities of these events. These are the statistically predictable unknowns.

The ideal tool for studying stationary signals is the Fourier transform. In other words, stationary signals decompose canonically into linear combinations of waves (sines and cosines). In the same way, signals that are not stationary decompose into linear combinations of wavelets.

The study of nonstationary signals, where transient events appear that cannot be predicted (even statistically with knowledge of the past), necessitates techniques different from Fourier analysis. These techniques, which are specific to the nonstationarity of the signal, include wavelets of the "time-frequency" type and wavelets of the "time-scale" type. "Time-frequency" wavelets are suited, most specifically, to the analysis of quasi-stationary signals, while "time-scale" wavelets are adapted to signals having a fractal structure.

Before defining "time-frequency" wavelets and "time-scale" wavelets, we indicate their common points. They belong to a more general class of algorithms that are encountered as often in mathematics as in speech processing. Mathematicians speak of "atomic decompositions," while speech specialists speak of "decompositions in time-frequency atoms"; *the scientific reality is the same in both cases.*

An "atomic decomposition" consists in extracting the simple constituents that make up a complicated mixture. However, contrary to what happens in chemistry, the "atoms" that are discovered in a signal will depend on the point of view adopted for the analysis. These "atoms" will be "time-frequency atoms" when we study quasi-stationary signals, but they could, in other situations, be replaced by "time-scale wavelets" or "Grossmann–Morlet wavelets."

These "atoms" or "wavelets" have no more physical existence than the number system used to multiply the mass of the earth by that of the moon. Each number system has an internal logical coherence, but no scientific law asserts that multiplication must of necessity be done in base 10 rather than base 2. On the other hand, the number system used by the Romans is certainly excluded because it is not suitable for multiplication.

Having different algorithms that allow us to code a signal by decomposing it into "time-frequency atoms" presents us with a similar situation. The decision to use one or the other of these algorithms will be made by considering their *performance*. How well they perform must be judged in terms of one of the anticipated *goals* of signal processing. *An algorithm that is optimal for compression can be disastrous for analysis: A standard energetic criterion for the compression could cause details that are important for the analysis to be systematically neglected.*

These thoughts will be developed and clarified in §§1.6 and 1.7. It is time to define wavelets, which we do in the next two sections.

1.4.   Grossmann–Morlet time-scale wavelets.

"Time-scale" analysis (which should be called "space-scale" in the image case, but which we prefer to call "multiresolution analysis") involves using a vast range of scales for signal analysis. This notion of scale, which clearly refers to cartography, implies that the signal (or image) is replaced, at a given scale, by the best possible approximation that can be drawn at that scale. By "traveling" from the large scales toward the fine scales, one "zooms in" and arrives at more and more exact representations of the given signal.

The analysis is then done by calculating the change from one scale to the next. These are the details that allow one, by correcting a rather crude approximation, to move toward a better quality representation. This algorithmic scheme is called "multiresolution analysis" and is developed in Chapters 3 and 4. *Multiresolution analysis is equivalent to an atomic decomposition where the atoms are Grossmann–Morlet wavelets.*

We define these wavelets by starting with a function $\psi(t)$ of the real variable t. This function is called a "mother wavelet" provided it is well localized and oscillating. (By oscillating it resembles a wave, but by being localized it is a wavelet.) The localization condition is expressed in the usual way as decreasing rapidly to zero when $|t|$ tends to infinity. The second condition suggests that $\psi(t)$ vibrates like a wave: Here we require that the integral of $\psi(t)$ be zero and that the same hold true for the first m movements of ψ. This is expressed as

$$(1.1) \qquad 0 = \int_{-\infty}^{\infty} \psi(t)\, dt = \cdots = \int_{-\infty}^{\infty} t^{m-1}\psi(t)\, dt.$$

The "mother wavelet," $\psi(t)$, generates the other wavelets, $\psi_{(a,b)}(t)$, $a > 0$, $b \in \mathbb{R}$, of the family by change of scale (the scale of $\psi(t)$ is conventionally 1, and that of $\psi_{(a,b)}(t)$ is $a > 0$) and translation in time (the function $\psi(t)$ is conventionally centered around 0, and $\psi_{(a,b)}(t)$ is then centered around b).

Thus we have

$$(1.2) \qquad \psi_{(a,b)}(t) = \frac{1}{\sqrt{a}} \psi\left(\frac{t-b}{a}\right), \qquad a > 0, \qquad b \in \mathbb{R}.$$

Alex Grossmann and Jean Morlet have shown that, if $\psi(t)$ is real-valued, this collection can be used as if it were an orthonormal basis. This means that any signal of finite energy can be represented as a linear combination of wavelets $\psi_{(a,b)}(t)$ and that the coefficients of this combination are, up to a normalizing factor, the scalar products $\int_{-\infty}^{\infty} f(t)\psi_{(a,b)}(t)\,dt$.

These scalar products measure, in a certain sense, the fluctuations of the signal $f(t)$ around the point b, at the scale given by $a > 0$.

It required uncommon scientific intuition to assert, as Grossmann and Morlet did, that this new method of time-scale analysis was suitable for the analysis and synthesis of *transient signals*. Signal processing experts were annoyed by the intrusion of these two poachers on their preserve and made fun of their claims.

This polemic died out after only a few years. In fact, the argument should never have arisen because the methods of time-scale or multiresolution analysis had existed for five or six years under various disguises: in signal analysis (under the name of quadrature mirror filters) and in image analysis (under the name of pyramid algorithms).

The first to report on this was Stephane Mallat. He constructed a guide that allowed the same signal analysis method to be recognized under very different presentations: wavelets, pyramid algorithms, quadrature mirror filters, Littlewood–Paley analysis, David Marr's zero crossings...

Mallat's brilliant synthesis has been the source of many new developments. One of the most notable of these is Ingrid Daubechies's discovery of orthonormal wavelet bases having preselected regularity and compact support. The only previously known case was the Haar system (1909), which is not regular. Thus 80 years separated Alfred Haar's work and its natural extension by Daubechies [1].

1.5.   Time-frequency wavelets from Gabor to Malvar.

Dennis Gabor was the first to introduce *time-frequency wavelets* or *Gabor wavelets*. He had the idea to divide a wave (whose mathematical representation is $\cos(\omega t + \varphi)$) into segments and then to throw away all but one of these segments. This left a piece of a wave, or a wavelet, which had a beginning and an end.

To use a musical analogy, a wave corresponds to a note (a *ré*, for example) that has been emitted since the origin of time and sounds indefinitely, without attenuation, until the end of time. A wavelet then corresponds to the same *ré* that is struck at a certain moment, say, on a piano, and is later muffled by the pedal. In other words, a Gabor wavelet has (at least) three pieces of information: a beginning, an end, and a specific frequency in between.

Difficulties appeared when it was necessary to decompose a signal using Gabor wavelets. As long as one does only continuous decompositions (using all frequencies and all time), Gabor wavelets can be used as if they formed an orthonormal basis. But the corresponding discrete algorithms do not exist, or they require so much tinkering that they become too complicated.

It is only very recently, by abandoning Gabor's approach, that two separate groups have discovered time-frequency wavelets having good algorithmic qualities. These special time-frequency wavelets, called Malvar wavelets, are particularly well suited for coding speech and music. Moreover, they are equally useful to the FBI for storing fingerprints.

The decomposition of a signal in an orthonormal basis of Malvar wavelets imitates writing music using a musical score. But this composition is misleading because a piece of music can be written in only one way, whereas there exists a nondenumerable infinity of orthonormal bases of Malvar wavelets. Choosing one among these is equivalent to segmenting the given signal and then doing a traditional Fourier analysis on the delimited pieces. What is the best way to choose this segmentation? This question leads us naturally to the following section.

1.6. Optimal algorithms in signal processing.

Which wavelet to choose? I have often posed this question at meetings on wavelets and their applications held since 1985. But this question needs to be sharpened. *What freedom of choice is at our disposal? What are the objectives of the choices we make? Can we make better use of the choices offered to us by considering the anticipated goals?* These are several of the questions we will try to answer.

The goal we have in mind is aptly illustrated by a remark by Benoît Mandelbrot made in an interview on "France-Culture": "The world around us is very complicated. The tools at our disposal to describe it are very weak."

It is notable that Mandelbrot used the word "describe" and not "explain" or "interpret." We are going to follow him in this, ostensibly, very modest approach. This is our answer to the problem about the objectives of the choices: *Wavelets, whether they are of the time-scale or time-frequency type, will not help us to explain scientific facts, but they will serve to describe the reality around us, whether or not it is scientific.*

Our task is to optimize the description. This means that we must make the best use of the resources allocated to us (for example, the number of available bits) to obtain the most precise possible description.

To resolve this problem, we must first indicate how the quality of the description will be judged. Most often, the criteria used are academic and do not correspond at all to the user's point of view.

For example, in image processing, all the calculations for judging the quality of the description use the quadratic mean value of gray levels. It is clear, however, that our eye has a much more selective sensitivity. Thus, in the last analysis, we should submit the performance of an "optimal algorithm" to the users because the average approximation criterion that leads to this algorithm will often be inadequate.

The case of speech (telephonic communication) or music is similar. After systematic research that optimizes the reception quality (quality calculated with an average), it would be advisable to experiment by taking into account the judgments of telephone users and musicians.

Thus we see a two-state program developing. Nevertheless, the only stage we describe in the following pages is the "objective search" for an optimal algorithm, even though its optimality is defined in terms of a debatable energy criterion.

Rather than formulate ad hoc algorithms for each signal or each class of signals, we are going to construct, once and for all, a vast collection called a library of algorithms. We will also construct a "meta-algorithm" whose function will be to find the particular algorithm in the library that best serves the given signal in view of the criterion for the quality of the description.

It is hardly an exaggeration to say that we will introduce almost as many analysis algorithms as there are signals. For example, for a signal recorded on $2^{10} = 1024$ points, the number of algorithms at our disposal will be of the order 2^{1024}. This is the number of all the signals defined on our 1024 points that take only two values, 0 and 1.

Thus we will use a "very large library" to describe the signals, but we exclude the "library of Babel." This would contain all the books, or all the signals in our case. And as everyone knows, the search for a specific book in the library of Babel is an insurmountable task. The "ideal library" must be sufficiently rich to suit all transient signals, but the "books" must be easily accessible.

While a single algorithm (Fourier analysis) is appropriate for all stationary signals, the transient signals are so rich and complex that a single analysis method (whether of time-scale or time-frequency) cannot serve them all.

If we stay in the relatively narrow environment of Grossmann–Morlet wavelets, also called time-scale algorithms, we have only two ways to adapt the algorithm to the signal being studied: We can choose one or another analyzing wavelet, and we can use either the continuous or the discrete version of the wavelets.

For example, we can require the analyzing wavelet ψ to be an analytic signal, which means that its Fourier transform $\hat{\psi}(\omega)$ is zero for negative frequencies. In this case, all the wavelets $\psi_{(a,b)}$, $a > 0$, $b \in \mathbb{R}$, generated by ψ will still have this property, and their linear combinations given by the algorithm will be the analytic signal F associated with the real signal f.

Likewise, we can follow Daubechies and, for a given $r \geq 1$, choose for $\psi(x)$ a real-valued function in the class C^r with compact support such that the collection $2^{j/2}\psi(2^j x - k)$, $j,k \in \mathbb{Z}$, is an orthonormal basis for $L^2(\mathbb{R})$. In this discrete version of the algorithm, $a = 2^{-j}$ and $b = k2^{-j}, j,k \in \mathbb{Z}$.

In spite of this, the choices that can be made from the set of time-scale wavelets remain limited. *The search for optimal algorithms leads us on some remarkable algorithmic adventures, where time-scale wavelets and time-frequency wavelets are in competition, and where they are also compared to intermediate algorithms that mix the two extreme forms of analysis.*

These considerations are developed in Chapters 6 and 7 and the question I asked myself six years ago (what wavelet to choose?) seems passé to me today. The choices that we can and must consider no longer involve only the analyzing instrument (the wavelet); they also involve the methodology employed (time-scale, time-frequency, or intermediate algorithms).

Today, the competing algorithms (time-scale and time-frequency) are included in a whole universe of intermediate algorithms. An entropy criterion permits us to choose the algorithm that optimizes the description of the given signal within the given bit-allocation.

Each algorithm is presented in terms of a particular orthogonal basis. We can compare searching for the optimal algorithm to searching for the best point of view, or best perspective, to look at a statue in a museum. Each point of view reveals certain parts of the statue and obscures others. We change our point of view to find the best one by going around the statue. In effect, we make a rotation; we change the orthonormal basis of reference to find the optimal basis.

These reflections lead us quite naturally to the scientific thoughts of David Marr, which we present in the next section.

1.7. Optimal representation, according to Marr.

David Marr was fascinated and intrigued by the complex relations that exist between the choice of a representation of a signal and the nature of the operations or transformations that such a representation permits. He wrote [5, pp. 20–21]:

A *representation* is a formal system for making explicit certain entities or types of information, together with a specification of how the system does this. And I shall call the result of using a representation to describe a given entity a description of the entity in that representation.

For example, the Arabic, Roman and binary numerical systems are all formal systems for representing numbers. The Arabic representation consists of a string of symbols drawn from the set $\{0, 1, 2, 3, 4, 5, 6, 7, 8, 9\}$ and the rule for constructing the description of a particular integer n is that one decomposes n into a sum of multiples of powers of 10... A musical score provides a way of representing a symphony; the alphabet allows the construction of a written representation of words...

A representation, therefore, is not a foreign idea at all—we all use representations all the time. However, the notion that one can capture some aspect of reality by making a description of it using a symbol and that to do so can be useful seems to me a fascinating and powerful idea. But even the simple examples we have discussed introduce some rather general and important issues that arise whenever one chooses to use one particular representation.

For example, if one chooses the Arabic numerical representation, it is easy to discover whether a number is a power of 10 but difficult to discover whether it is a power of 2. If one chooses the binary representation, the situation is reversed.

Thus, there is a trade-off; *any particular representation makes certain informa-tion explicit at the expense of information that is pushed into the background and may be quite hard to recover.*[1]

This issue is important, because how information is presented can greatly affect how easy it is to do different things with it. This is evident even from our numbers example: it is easy to add, to subtract and even to multiply if the Arabic or binary representations are used, but it is not at all easy to do these things—especially multiplication—with Roman numerals. This is a key reason why the Roman culture failed to develop mathematics in the way the earlier Arabic cultures had...

There is an essential difference between Marr's considerations and the algorithms that we develop in the first six chapters. The difference is that the choice of the best representation, according to Marr, is tied to an objective goal. For the problem posed by vision, the goal is to extract the contours, recognize the edges of objects, delimit them, and understand their three-dimensional organization. In the algorithms we develop, the only criterion is internal to the algorithm and consists in reducing the amount of data. We have not yet studied situations where one must also take into consideration an external criterion. In spite of this difference, Marr's point of view is very close to the one we will present.

1.8.   Terminology.

The elementary constituents used for signal analysis and synthesis will be called, depending on the circumstances, wavelets, time-frequency atoms, or wavelet packets (Chapter 8).

The wavelets used will be either Grossmann–Morlet wavelets of the form

$$\frac{1}{\sqrt{a}}\,\psi\left(\frac{t-b}{a}\right), \qquad a > 0, \qquad b \in \mathbb{R},$$

the wavelets of Daubechies that have the form

$$2^{j/2}\psi(2^j t - k), \qquad j, k \in \mathbb{Z},$$

or Gabor–Malvar wavelets of the form

$$\omega(t - l)\cos[\pi(k + 1/2)(t - l)], \qquad k \in \mathbb{N}, \qquad l \in \mathbb{Z}.$$

In the first two cases, we will speak of time-scale algorithms; in the last case, we will speak of time-frequency algorithms. Later we will mix the two points of view and subject the Gabor–Malvar wavelets to dyadic dilations to construct the Daubechies wavelets. One thus encounters generalized time-frequency atoms.

We will use only two "very large libraries." The first consists of orthonormal bases whose elements are wavelet packets. In the second, the wavelet packets are replaced by the generalized time-frequency atoms that we have just described.

---
[1]The italics are ours.

1.9. Reader's guide.

In Chapters 2 through 7, we present the time-scale algorithms (Chapters 3 and 4) and time-frequency algorithms (Chapters 5, 6, and 7). Chapter 2 has a special status. We have tried to retrace the path that led from Fourier analysis (Fourier 1807) to wavelet analysis (Calderón 1960 and Strömberg 1981) and to the very core of contemporary mathematics.

Quadrature mirror filters are studied in Chapter 3 in connection with problems posed by the digital telephone.

The pyramid algorithms described in Chapter 4 concern numerical image processing. The orthogonal pyramids use precisely the quadrature mirror filters of Chapter 3. The pyramid algorithms lead either to orthogonal wavelets or to bi-orthogonal wavelets.

From Chapter 5 on, we will study time-frequency algorithms. The Wigner–Ville transform enables "the signal to be displayed in the time-frequency plane." After indicating the main properties of the Wigner–Ville transform, we must point out that it does not provide an algorithm that allows us to decompose a signal into time-frequency atoms that are the analogues of musical notes.

The two algorithms that do provide access to these "atomic decompositions" are presented in Chapters 6 (Malvar wavelets) and 7 (wavelet packets).

The first six chapters form a coherent unit. This is not the case for the last four chapters; each of them treats a special application of wavelets and time-scale methods. Chapter 8 deals with the Marr–Mallat theory. This concerns the possibility of coding an image using the zero-crossings of its wavelet transform.

Using the Grossmann–Morlet analysis one can determine the fractal exponents, as a function of position, of a multifractal object. This is one of the most remarkable applications of the Grossmann–Morlet wavelets, and it is presented in Chapter 6.

The last two chapters cover two very dynamic subjects: turbulence and the multifractal approach to turbulence (Chapter 10) and the hierarchical organization of the galaxies and the structure of the Universe (Chapter 11). In both cases, results are still meager. But these avenues of research are fascinating, and the researchers involved have such good reputations that one can reasonably expect important progress before long.

Bibliography

[1] I. DAUBECHIES, *Ten Lectures on Wavelets*, Society for Industrial and Applied Mathematics, Philadelphia, PA, 1992.

[2] D. GABOR, *Theory of communication*, J. IEE, 93 (1946), pp. 429–457.

[3] C. GASQUET AND P. WITOMSKI, *Analyse de Fourier et applications, Filtrage, Calcul numérique, Ondelettes*, Masson, Paris, 1990.

[4] J. L. LIONS, *La planète terre: rôle des mathèmatiques et des super-ordinateurs*, cours à l'Institut d'Espagne, 1990.

[5] D. MARR, *Vision*, W. H. Freeman and Co., New York, 1982.

[6] R. VAUTARD AND M. GHIL, *Singular spectrum analysis in nonlinear dynamics, with applications to paleoclimatic time series*, Physica D, 35 (1989), pp. 359–424.

Wavelets From a Historical Perspective

2.1. Introduction.

The application of wavelets to signal and image processing is only a few years old. But in looking back over the history of mathematics, we will uncover at least seven different origins of wavelet analysis. Most of this work was done around the 1930s, and, at that time, the separate efforts did not appear to be parts of a coherent theory. In particular, neither the word "wavelet" nor the corresponding concept appeared in this research done a half-century ago. Only today do we know that this work prefigured the theory of wavelets.

It is important to describe these seven sources in some detail. Each of them corresponds to a specific point of view and a particular technique, which, only now, are we able to view from a common scientific perspective. What's more, these specific techniques were rediscovered several years ago by physicists and mathematicians working on wavelets. For example, the Littlewood–Paley analysis (§2.4), which dates from 1930, underlies Mallat's work on image processing. Matthias Holschneider used, without knowing it, Lusin's technique (1930) to clarify the fractal structure of Riemann's function (§2.6 and §9.4). Grossmann and Morlet rediscovered Alberto Calderón's identity (1960) 20 years later. And, to spare no one, the author of these lines was not the first to construct a regular, well-localized orthonormal wavelet basis having the algorithmic structure of Haar's system (1909); J. O. Strömberg had done the same thing five years earlier.

Does this mean that everything had already been written and that "team-wavelet" researchers appropriated—while making a great show of it—results discovered by others? This judgment must be qualified for two reasons. In the first place, the physicists from "team wavelet" were not aware of Calderón's work; it was in completely good faith that they presented as a revolutionary innovation results that were about 20 years old. "Team wavelet" researchers can be accused of ignorance but not of plagiarism. But above all, by rediscovering these known scientific facts, the investigators gave them new life and authority. Our debt to Grossmann and Morlet is not so much for having rediscovered the identity of Calderón as it is for having related it to processing nonstationary signals. This bold synthesis certainly encountered resistance, and Calderón himself found this use of his work incongruous.

The role of Morlet and Grossmann in the wavelet saga can be compared to that of Mandelbrot in the story of fractals. It is true that before Mandelbrot there were Pierre Fatou and Gaston Julia, and certain of Mandelbrot's discoveries repeat their work. But Mandelbrot showed us the possibility of interpreting the world around us with the help of a new concept, that of fractals. None of the mathematicians working on Hausdorff dimension were ready, from their training or experience, for such a leap into the unknown. Wavelets have gone the same way, and one of our objectives is to construct the bridge that relates signal processing to the different mathematical efforts that developed outside the "theory of wavelets."

2.2. From Fourier (1807) to Haar (1909), frequency analysis becomes scale analysis.

Let's go back to the origins, that is to Joseph Fourier. As everyone knows, he asserted in 1807 that any 2π-periodic function $f(x)$ is the sum $a_0 + \sum_1^\infty (a_k \cos kx + b_k \sin kx)$ of its "Fourier series." The coefficients a_0, a_k, and $b_k (k \geq 1)$ are calculated by

$$a_0 = \frac{1}{2\pi} \int_0^{2\pi} f(x)\, dx,$$

and by

$$a_k = \frac{1}{\pi} \int_0^{2\pi} f(x) \cos kx\, dx, \qquad b_k = \frac{1}{\pi} \int_0^{2\pi} f(x) \sin kx\, dx.$$

When Fourier announced his surprising results, neither the notion of function nor that of integral had yet received a precise definition. We can even say that Fourier's statement played an essential role in the evolution of the ideas mathematicians have had about these concepts.

Before Fourier's work, entire series were used to represent and manipulate functions, and the most general functions that could be constructed were endowed with very special properties indeed. Furthermore, these properties were unconsciously associated with the notion of function itself. By passing from a representation of the form

(2.1) $a_0 + a_1 x + a_2 x^2 + a_3 x^3 + \ldots$

to one of the form

(2.2) $a_0 + (a_1 \cos x + b_1 \sin x) + (a_2 \cos 2x + b_2 \sin 2x) + \ldots$

Fourier discovered, without knowing it, a new functional universe.

However, in 1873, Paul Du Bois-Reymond constructed a continuous, 2π-periodic function of the real variable x, whose Fourier series diverged at a given point. If Fourier's assertion were true, it could not be so in the naïve sense imagined by Fourier.

At that time, three new avenues were opened to mathematicians, and all three have led to important results:

(a) They could modify the notion of function and find one that is adapted, in a certain sense, to Fourier series;

(b) They could modify the definition of convergence of Fourier series; or

(c) They could find other orthogonal systems for which the phenomenon, discovered by Du Bois-Reymond in the case of the trigonometric system, cannot happen.

The functional concept best suited to Fourier series was created by Henri Lebesgue. It involves the space $L^2[0, 2\pi]$ of (classes of) functions that are square-integrable on the interval $[0, 2\pi]$. The sequence

$$\frac{1}{\sqrt{2\pi}}, \quad \frac{1}{\sqrt{\pi}} \cos x, \quad \frac{1}{\sqrt{\pi}} \sin x, \quad \frac{1}{\sqrt{\pi}} \cos 2x, \quad \frac{1}{\sqrt{\pi}} \sin 2x, \dots$$

is an orthonormal basis for this space. Furthermore, the coefficients of the decomposition in this orthonormal basis form a square-summable series, and this expresses the conservation of energy: The quadratic mean value of the developed function $f(x)$ is (up to a normalization factor) the sum of the squares of the coefficients. Finally, the Fourier series of f converges to f in the sense of the quadratic mean.

The second way that was followed to avoid the difficulty raised by Du Bois-Reymond was to modify the notion of convergence. The partial sums $S_n(x)$ are replaced by the Cesàro sums $\sigma_n = \frac{1}{n}(S_0 + \cdots + S_{n-1})$, and everything falls into place.

The third route leads to wavelets. This was followed by Haar, who asked himself this question: "Does there exist another orthonormal system $h_0(x), h_1(x), \dots, h_n(x), \dots$ offunctions defined on [0,1] such that for any function $f(x)$, continuous on [0,1], the series

$$(2.3) \qquad \langle f, h_0 \rangle h_0(x) + \langle f, h_1 \rangle h_1(x) + \cdots + \langle f, h_n \rangle h_n(x) + \dots$$

converges to $f(x)$ uniformly on [0,1]?"

Here we have written $\langle u, v \rangle = \int_0^1 u(x)\overline{v(x)}\, dx$, where $\overline{v}$ is the complex conjugate of v, and chosen the interval [0,1] for convenience to fix our ideas.

As we will see, this problem has an infinite number of solutions. In 1909, Haar discovered the simplest solution and, at the same time, opened one of the routes leading to wavelets.

Haar begins with the function $h(x)$ that is equal to 1 on $[0,1/2)$, -1 on $[1/2,1)$, and 0 outside the interval $[0,1)$. For $n \geq 1$, he writes $n = 2^j + k$, $j \geq 0$, $0 \leq k < 2^j$, and defines $h_n(x) = 2^{j/2} h(2^j x - k)$. The support of $h_n(x)$ is the dyadic interval $I_n = [k2^{-j}, (k+1)2^{-j})$, which is included in $[0,1)$ when $0 \leq k < 2^j$. To complete the set, define $h_0(x) = 1$ on $[0,1)$. Then the series $h_0(x), h_1(x), \dots, h_n(x), \dots$ is an orthonormal basis (also called a Hilbert basis) for $L^2[0,1]$.

The uniform approximation of $f(x)$ by the sequence $S_n(f)(x) = \langle f, h_0 \rangle h_0(x) + \cdots + \langle f, h_n \rangle h_n(x)$ is nothing more than the classical approximation of a continuous function by step functions whose values are the mean values of $f(x)$ on the appropriate dyadic intervals.

We can criticize the Haar construction on a couple of points. On one hand, the "atoms" $h_n(x)$ used to construct the continuous function $f(x)$ are not themselves continuous functions, and thus there is a lack of coherence.

But there is a more serious criticism. Suppose that, instead of being continuous on the interval $[0,1]$, $f(x)$ is a function of the class C^1, which means $f(x)$ is continuous and has a continuous derivative. Then the approximation of $f(x)$ by step functions would be completely inappropriate. In this case, a suitable approximation would be the one created from the graph of $f(x)$ by inscribing polygonal lines.

The Haar construction is suitable only for continuous functions, functions square-integrable on $[0,1]$ or, more generally, functions whose index of regularity is near 0. We will see a little later what this means.

These two defects of the Haar system and the idea of approximating the graph of $f(x)$ with inscribed polygonal lines led Faber and Schauder to replace the functions $h_n(x)$ of the Haar system by their primitives. This research began in 1910 and continued until 1920.

Define the "triangle function" $\Delta(x)$ by $\Delta(x) = 0$ if $x \notin [0,1]$, $\Delta(x) = 2x$ if $0 \le x \le 1/2$, and $\Delta(x) = 2(1-x)$ if $1/2 \le x \le 1$. Then Faber and Schauder considered the series $\Delta_n(x)$, $n \ge 1$, defined by

$$(2.4) \quad \Delta_n(x) = \Delta(2^j x - k) \quad \text{for} \quad n = 2^j + k, \qquad j \ge 0, \qquad 0 \le k < 2^j.$$

The support of $\Delta_n(x)$ is the dyadic interval $I_n = [k2^{-j}, (k+1)2^{-j}]$, and $\Delta_n(x)$ is the primitive of $h_n(x)$ multiplied by $2 \cdot 2^{j/2}$ and zero outside I_n.

For $n = 0$, we set $\Delta_0(x) = x$, and we add the constant 1 to complete the set of functions. Then the sequence $1, \Delta_0(x), \ldots, \Delta_n(x), \ldots$ is a *Schauder basis* for the Banach space E of continuous functions on $[0,1]$. This means that every continuous function on $[0,1]$ can be written as

$$(2.5) \qquad\qquad f(x) = a + bx + \sum_{1}^{\infty} \alpha_n \Delta_n(x)$$

and that the series has the following properties: the convergence is uniform on $[0,1]$ and, as a consequence, the coefficients are unique.

We note that the Haar system is not a Schauder basis of E because a Schauder basis of a Banach space must be made up of vectors of that space, and the functions h_n are not continuous.

The calculation of the coefficients in (2.5) is immediate. We have successively $f(0) = a$ and $f(1) = a + b$, which gives a and b. This allows us to consider a function $f(x) - a - bx$, which is zero at 0 and 1. Once this reduction is made, we have $f(1/2) = \alpha_1$, which allows us to consider a function equal to zero at 0, $1/2$, and 1. The calculation continues with $f(1/4) = \alpha_2$ and $f(3/4) = \alpha_3$, and so on. If we do not wish to "peel" $f(x)$ this way, the coefficients α_n can be computed directly by the formula

$$(2.6) \qquad \alpha_n = f((k+1/2)2^{-j}) - \frac{1}{2}[f(k2^{-j}) + f((k+1)2^{-j})].$$

We can give a further interpretation to (2.5). If, instead of being continuous, $f(x)$ was a function in the class C^1, then we could differentiate (2.5) term by term and obtain the expansion of $f'(x)$ in the Haar basis. If $f(x)$ is in class C^1, the series (2.5) converges uniformly to $f(x)$ and the series differentiated term by term converges uniformly to $f'(x)$. Does this mean that the functions $\Delta_n(x)$, $n \geq 0$, with the added function 1, constitute a Schauder basis for the Banach space $C^1[0,1]$? As before, this is not the case because the functions $\Delta_n(x)$ do not belong to the space in question.

Following Hölder, we define the space $C^r[0,1]$, for $0 < r < 1$, by the relation $|f(x) - f(y)| \leq C|x - y|^r$ for a certain constant C and for all $x, y \in [0,1]$. Then it is clear from (2.6) that $|\alpha_n| \leq C2^{-(j+1)r}$ if f belongs to C^r. Since $2^j \leq n < 2^{j+1}$ we can also write $|\alpha_n| \leq Cn^{-r}$, $n \geq 1$. The converse, although much less evident, is nevertheless true when $0 < r < 1$. It is not true if $r = 1$.

Contemporary physicists are very interested in the Hölder spaces C^r because they occur naturally in the study of fractal structures. In fact, the physicists require more. They wish to calculate a Hölder exponent r that varies from one point to another. Here is the definition of pointwise Hölder exponents. We say that $f(x)$ satisfies a Hölder condition of exponent r, $0 < r < 1$, at x_0 if $|f(x) - f(x_0)| \leq C|x - x_0|^r$. Then we look for the largest possible value of r, which is denoted $r(x_0)$ if it exists.

Contemporary science deals with numerous physical phenomena having multifractal structures. This means that the "fractal exponents" $r(x_0)$ vary from point to point. In this case, the Hausdorff dimension of the set of points x_0, where $r(x_0) = \alpha$ is a function of $d(\alpha)$ whose graph has a distinctive form.

An example from mathematics is the celebrated function attributed to Bernhard Riemann $\sum_1^\infty (1/n^2) \sin(n^2 x)$. This example illustrates the point that the Fourier series of a function provides no directly accessible information about the function's multifractal structure. By using the wavelets of Lusin (which we present in §2.6), it is possible to elucidate the multifractal structure of Riemann's function. This was done by Matthias Holschneider and Philippe Tchamitchian. We describe their work in Chapter 9.

A second example is the signal coming from fully developed turbulence. The multifractal structure of this signal has been studied with great care by Alain Arnéodo and his collaborators. We present this example in Chapter 10.

Conceivably, the pointwise Hölder exponents could be computed by going back to the definition. However, the example of the Riemann functions shows that such an approach is too crude to yield practical results. This approach offers no way to take into consideration the inevitable noise (assumed to be Gaussian) that alters a signal. The Schauder basis presents the same difficulties because the calculation of the coefficients α_n (according to (2.6)) calls directly upon explicit values of the signal.

Today, we are fortunate to have much more subtle ways to attack this problem. Specifically, the pointwise Hölder exponents are now determined using wavelet analysis. The wavelet coefficients replace those given by formula (2.6). They are less sensitive to noise because they measure, at different scales, the average fluctuations of the signal. These methods will be described in Chapter 9.

2.3. New directions of the 1930s: Lévy and Brownian motion.

Brownian motion is a random signal. We will limit our discussion to the one-dimensional case. We thus write $X(t, \omega)$ for the Brownian motion: t denotes time, ω belongs to a probability space Ω, and $X(t, \omega)$ is regarded as a real-valued function of time depending on the parameter ω.

To obtain a realization of Brownian motion, we choose a particular orthonormal basis $Z_i(t)$, $i \in I$, for the usual Hilbert space $L^2(\mathbb{R})$. Then we know that the derivative (in the sense of distributions) $\frac{d}{dt} X(t, \omega)$ is written as

$$(2.7) \qquad \frac{d}{dt} X(t, \omega) = \sum_{i \in I} g_i(\omega) Z_i(t),$$

where the $g_i(\omega)$, $i \in I$, are independent identically distributed Gaussian random variables with zero mean.

Then the problem is to choose the best possible representation of Brownian motion. It is certainly advisable, as in all signal processing problems, to have in mind the desired end result.

If we wish to examine the spectral properties of Brownian motion, we are led to select the Fourier representation. The real line is cut into intervals $[2l\pi, 2(l+1)\pi]$, $l \in \mathbb{Z}$, and the trigonometric system is used on each of the intervals. In its real form, this trigonometric system is $\frac{1}{\sqrt{2\pi}}$, $\frac{1}{\sqrt{\pi}} \cos kx$, and $\frac{1}{\sqrt{\pi}} \sin kx$, $k \geq 1$.

If we wish to highlight the multifractal structure of Brownian motion, Fourier analysis is inadequate. On the other hand, the analysis using the Schauder basis immediately reveals the Hölder regularity C^r, $r < 1/2$, of the Brownian motion trajectories.

We start with the Haar basis for $L^2(\mathbb{R})$ composed of the functions $h_n(t - l)$, $n \geq 0$, $l \in \mathbb{Z}$, and expand the white noise $\frac{d}{dt} x(t, \omega)$ in this orthonormal basis. By taking primitives, we obtain the development of Brownian motion in the Schauder basis.

To fix our ideas, we restrict the discussion to Brownian motion on the interval $[0,1]$. For this $l = 0$, and

$$(2.8) \qquad X(t, \omega) = a_0(\omega) + t b_0(\omega) + \frac{1}{2} \sum_1^{\infty} 2^{-j/2} g_n(\omega) \Delta_n(t),$$

where the $g_n(\omega)$ are independent Gaussian random variables with mean zero and the same distribution.

To verify that the function $X(t, \omega)$ belongs to the Hölder space C^r for almost all $\omega \in \Omega$, it is sufficient to show that $2^{-j/2} |g_n(\omega)| \leq C(\omega) 2^{-jr}$. If, for almost all $\omega \in \Omega$, one had $\sup_{n \geq 0} |g_n(\omega)| < \infty$, then the trajectories of the Brownian motion would almost surely belong to the space $C^{1/2}$. But this is not the case, and instead we have $\sup_{n \geq 2} (|g_n(\omega)| / \sqrt{\log n}) < \infty$ for almost all $\omega \in \Omega$. Then the criterion for Hölder regularity gives

$$(2.9) \qquad |X(t + h, \omega) - X(t, \omega)| \leq C(\omega) \sqrt{h \log 1/h},$$

where $C(\omega) < \infty$ for almost all $\omega \in \Omega$.

We see, from this theorem of Paul Lévy, the superiority of the Schauder basis over the Fourier basis for studying local regularity properties.

The determination of the fractal exponents requires some extensions. These extensions that allow the study of small, complicated details form part of "multiresolution analysis," which we define in the following chapters. These ideas are already incorporated in the definition of the Schauder basis itself, in particular, within the mapping $x \mapsto 2^j x - k$. They are clearly absent in the trigonometric system.

Patrick Flandrin [5] has extended this work to the case of fractional Brownian motion, as it was proposed by Benoît Mandelbrot and John W. van Ness to model certain noise. Albert Benassi, Stéphane Jaffard, and Daniel Roux [3] have generalized these ideas to certain Gaussian–Markov fields. All of this work demonstrates that multiresolution methods are adapted to the analysis and synthesis of these processes.

2.4. New directions of the 1930s: Littlewood and Paley.

We have shown, with the example of Brownian motion, that the trigonometric system does not provide direct and easy access to local regularity properties and that these properties are clearly apparent when examined with other representations.

Similar difficulties are encountered when we try to localize the energy of a function. To be more precise, the integral $\frac{1}{2\pi} \int_0^{2\pi} |f(x)|^2 dx$, which is the mean value of the energy, is given directly by the sum of the squares of the Fourier coefficients. However, it is often important to know if the energy is concentrated around a few points or if it is distributed over the whole interval $[0, 2\pi]$. This determination can be made by calculating $\frac{1}{2\pi} \int_0^{2\pi} |f(x)|^4 dx$, or more generally, $\frac{1}{2\pi} \int_0^{2\pi} |f(x)|^p dx$ for $2 < p < \infty$. When the energy is concentrated around a few points, this integral will be much larger than the mean value of the energy, while it will be the same order of magnitude when the energy is evenly distributed. We write $\|f\|_p = (\frac{1}{2\pi} \int_0^{2\pi} |f(x)|^p dx)^{1/p}$ and, for obvious reasons of homogeneity, we compare the norms $\|f\|_p$ to determine if the energy is concentrated or dispersed. But if p is different from 2, we can neither calculate nor even estimate these norms $\|f\|_p$ by examining the Fourier coefficients of f.

The information needed for this calculation is hidden in the Fourier series of f; to reveal it, it is necessary to subject the series to manipulations that were discovered by Littlewood and Paley as long ago as 1930.

Littlewood and Paley define the "dyadic blocks" $\Delta_j f(x)$ by

$$(2.10) \qquad \Delta_j f(x) = \sum_{2^j \leq k < 2^{j+1}} (a_k \cos kx + b_k \sin kx),$$

where $a_0 + \sum_1^\infty (a_k \cos kx + b_x \sin kx)$ denotes the Fourier series of f. Then

$$f(x) = a_0 + \sum_0^\infty \Delta_j f(x),$$

and the fundamental result of Littlewood and Paley is that there exists, for $1 < p < \infty$, two constants $C_p \geq c_p > 0$ such that

$$(2.11) \qquad c_p \|f\|_p \leq \left\| \left(|a_0|^2 + \sum_0^\infty |\Delta_j f(x)|^2 \right)^{1/2} \right\|_p \leq C_p \|f\|_p.$$

If $p = 2$, $C_p = c_p = 1$, and there is equality in (2.11).

Up to this point, wavelets have not yet appeared. The path that leads from the work of Littlewood and Paley to wavelet analysis passes through the research done by Antoni Zygmund's group at the University of Chicago. Zygmund and the mathematicians around him sought to extend to n-dimensional Euclidean space the results obtained in the one-dimensional periodic case by Littlewood and Paley.

It was at this point that the "mother wavelet" $\psi(x)$ appeared. It is an infinitely differentiable, rapidly decreasing function of x, defined on the Euclidean space $\mathbb{R}^n$, whose Fourier transform $\hat{\psi}(\xi)$ satisfies the following four conditions:

$$(2.12) \qquad\qquad \hat{\psi}(\xi) = 1 \quad \text{if} \quad 1 + \alpha \leq |\xi| \leq 2 - 2\alpha,$$

where α, is by hypothesis, chosen in the interval $(0,1/3]$,

$$(2.13) \qquad\qquad \hat{\psi}(\xi) = 0 \quad \text{if} \quad |\xi| \leq 1 - \alpha \quad \text{or} \quad |\xi| \geq 2 + 2\alpha,$$

$$(2.14) \qquad\qquad \hat{\psi}(\xi) \quad \text{is infinitely differentiable on} \quad \mathbb{R}^n,$$

and

$$(2.15) \qquad\qquad \sum_{-\infty}^\infty |\hat{\psi}(2^{-j}\xi)|^2 = 1 \quad \text{for all} \quad \xi \neq 0.$$

Condition (2.15) does not amount to much. It is sufficient to verify it for $1 - \alpha \leq |\xi| \leq 2 - 2\alpha$, and then only two cases arise: if $1 - \alpha \leq |\xi| \leq 1 + \alpha$, (2.15) reduces to $|\hat{\psi}(\xi)|^2 + |\hat{\psi}(2\xi)|^2 = 1$, while if $1 + \alpha \leq |\xi| \leq 2 - 2\alpha$, (2.15) is automatically satisfied since one term is equal to 1 and all the others are zero.

Condition (2.15) implies that the analysis of Littlewood–Paley–Stein (whose definition will be given in a moment) conserves energy. This same condition (2.15) is satisfied by all the orthonormal wavelet bases of the form $2^{j/2}\psi(2^j x - k)$, $j, k \in \mathbb{Z}$. It also anticipates similar conditions shared by the quadrature mirror filters (Chapter 3) and the Malvar wavelets (Chapter 6).

The theory for $\mathbb{R}^n$ proceeds by setting $\psi_j(x) = 2^{nj}\psi(2^j x)$ and replacing the dyadic blocks of Littlewood and Paley with the convolution products $\Delta_j(f) = f * \psi_j$. The "Littlewood–Paley–Stein function" is defined by

$$g(x) = \left(\sum_{-\infty}^\infty |\Delta_j(f)(x)|^2 \right)^{1/2}.$$

If $f(x)$ belongs to $L^2(\mathbb{R}^n)$, the same is true for $g(x)$, and $\|f\|_2 = \|g\|_2$ (the conservation of energy).

If $1 < p < \infty$, there exist two constants $C_p \geq c_p > 0$ such that, for all functions f belonging to $L^p(\mathbb{R}^n)$,

(2.16) $$c_p\|g\|_p \leq \|f\|_p \leq C_p\|g\|_p,$$

where

$$\|f\|_p = \left(\int_{\mathbb{R}^n} |f(x)|^p dx\right)^{1/p}.$$

This double inequality (2.16) does not hold in the limiting case where $\alpha = 0$ [4].

The Littlewood–Paley–Stein function $g(x)$ provides a method for analyzing $f(x)$ in which a major role is played by the ability to vary arbitrarily the scales used in the analysis; by the same token, the notion of frequency plays a minor role. The dilations of size 2^j are present in the definition of the operators Δ_j. Nevertheless, conditions (2.12) and (2.13) endow these operators with a frequential content. *The sequence of operators Δ_j, $j \in \mathbb{Z}$, constitutes a bank of band-pass filters, oriented on frequency intervals covering approximately an octave.*

Thanks to the work of Marr and Mallat (which we describe in Chapter 8) the Littlewood–Paley analysis provides an effective algorithm for numerical image processing.

2.5. New directions of the 1930s: The Franklin system.

In 1927, Philip Franklin, who was a professor at the Massachusetts Institute of Technology (MIT), had the idea to create an orthonormal basis from the Schauder basis by using the Gram–Schmidt process. This gives a sequence $f_n(x)$ with $f_{-1}(x) = 1$, $f_0(x) = 2\sqrt{3}(x - 1/2), \ldots$, which is an orthonormal basis for $L^2[0,1]$. This sequence $(f_n)_{n \geq -1}$ is called the Franklin system and satisfies

(2.17) $$\int_0^1 f_n(x)dx = \int_0^1 x f_n(x)dx = 0 \quad \text{for} \quad n \geq 1.$$

The Franklin system has advantages of both the Haar basis and the Schauder basis. It can be used to decompose any function f in $L^2[0,1]$, which the Schauder basis does not allow, and it can be used to characterize the spaces C^r, $0 < r < 1$, by $|\langle f, f_n \rangle| \leq Cn^{-1/2-r}$, which the Haar basis does not allow. Thus the Franklin system works as well in relatively regular situations as it does in relatively irregular situations.

The weakness of the Franklin basis is that it no longer has a simple algorithmic structure. The functions of the Franklin basis, unlike those of the Haar basis or those of the Schauder basis, are not derived from a fixed function ψ by integer translations and dyadic dilations. This defect caused the Franklin system to be abandoned and forgotten for almost 40 years.

Fortunately, the Franklin system has survived this disgrace. Zbigniew Ciesielski revived it in 1963 by showing the existence of an exponent $\gamma > 0$ and a constant $C > 0$ such that

$$(2.18) \qquad |f_n(x)| \leq C2^{j/2} \exp(-\gamma|2^j x - k|)$$

if $0 \leq x \leq 1$, $n = 2^j + k$, $0 \leq k < 2^j$, and

$$(2.19) \qquad \left|\frac{d}{dx} f_n(x)\right| \leq C2^{3j/2} \exp(-\gamma|2^j x - k|).$$

Thus, everything works as if $f_n(x) = 2^{j/2}\psi(2^j x - k)$ and $\psi(x)$ were a Lipschitz function with exponential decay.

Today we have an asymptotic estimate for the functions $f_n(x)$. This estimate shows that, in a certain sense, the orthonormal wavelet basis discovered by Strömberg in 1980 was living hidden inside the Franklin system. We have, in fact, for $n = 2^j + k$, $0 \leq k < 2^j$,

$$(2.20) \qquad f_n(x) = 2^{j/2}\psi(2^j x - k) + r_n(x)$$

where, for a certain constant C,

$$(2.21) \qquad \|r_n(x)\|_2 \leq C(2 - \sqrt{3})^{d(n)}, \qquad d(n) = \inf(k, 2^j - k).$$

The function $\psi(x)$, which was discovered in 1980 by Strömberg, is completely explicit. It has the following three properties:

$$(2.22) \qquad \begin{array}{l} \psi(x) \text{ is continuous on the whole real line,} \\ \text{is linear on the intervals } [1,2], [2,3], \ldots, [l, l+1], \ldots \\ \text{and similarly on the intervals} \\ [1/2, 1], [0, 1/2], [-1/2, 0], \ldots, \left[-\dfrac{l+1}{2}, -\dfrac{l}{2}\right], \ldots, \end{array}$$

$$(2.23) \qquad |\psi(x)| \leq C(2 - \sqrt{3})^{|x|}, \quad \text{and}$$

$(2.24) \quad 2^{j/2}\psi(2^j x - k)$, $j, k \in \mathbb{Z}$, is an orthonormal basis for $L^2(\mathbb{R})$.

Note that $(2 - \sqrt{3}) < 1$, and hence (2.23) means that ψ decreases rapidly at infinity.

2.6. New directions in the 1930s: The wavelets of Lusin.

The interpretation of Lusin's work in terms of the theory of wavelets would probably astonish its author. But it is certainly the best reading, the one that gives the greatest beauty to Lusin's work.

We begin by introducing the object of Lusin's study, namely, the Hardy spaces $H^p(\mathbb{R})$, where $1 \leq p \leq \infty$. Let P denote the open, upper-half plane defined by $z = x + iy$ and $y > 0$. Then a function $f(x + iy)$ belongs to $H^p(\mathbb{R})$ if it is holomorphic in the half-plane P and if

$$(2.25) \qquad \sup_{y>0} \left(\int_{-\infty}^{\infty} |f(x+iy)|^p dx \right)^{1/p} < \infty.$$

When this condition is satisfied, the upper bound, taken over $y > 0$, is also the limit as y tends to 0. Furthermore, $f(x+iy)$ converges to a function denoted by $f(x)$ when y tends to 0, where convergence is in the sense of the L^p-norm. The space $H^p(\mathbb{R})$ can thus be identified with a closed subspace of $L^p(\mathbb{R})$, which explains the notation.

The Hardy spaces play a fundamental role in signal processing. One associates with a real signal $f(t)$, $-\infty < t < \infty$, of finite energy, the analytic signal $F(t)$ for which $f(t)$, $-\infty < t < \infty$, is the real part. By hypothesis, the energy of f is $\int_{-\infty}^{\infty} |f(t)|^2 dt$, and we require that $F(t)$ have finite energy as well. This implies that F belongs to the Hardy space $H^2(\mathbb{R})$. Then $F(t) = f(t) + ig(t)$, and the function $g(t)$ is the Hilbert transform of $f(t)$. For further information about analytic signals, the reader may refer to [9, pp. 118–119]. One may also consult the remarkable exposition by Jean Ville, [10].

Read in the light of the theory of wavelets, Lusin's work concerns the analysis and synthesis of functions in $H^p(\mathbb{R})$ using "atoms" or "basis elements," which are the elementary functions of $H^p(\mathbb{R})$. These are, in fact, the functions $(z - \bar{\zeta})^{-2}$, where the parameter ζ belongs to P.

Thus one wishes to obtain effective and robust representations of the functions $f(z)$ in $H^p(\mathbb{R})$ of the form

$$(2.26) \qquad f(z) = \iint_P (z - \bar{\zeta})^{-2} \alpha(\zeta) du\, dv,$$

where $\zeta = u + iv$ and where $\alpha(\zeta)$ plays the role of the coefficients. These coefficients should be simple to calculate, and their order of magnitude should provide an estimate of the norm of f in $H^p(\mathbb{R})$.

The *synthesis* is obtained by the following rule. We start with an *arbitrary* measurable function $\alpha(\zeta)$, subject only to the following condition introduced by Lusin: The quadratic functional $A(x)$ must be such that $\int_{-\infty}^{\infty} (A(x))^p dx$ is finite. This quadratic functional is defined by

$$(2.27) \qquad A(x) = \left(\iint_{\Gamma(x)} |\alpha(u + iv)|^2 v^{-2} du\, dv \right)^{1/2},$$

where

$$\Gamma(x) = \{(u, v) \in \mathbb{R}^2, \quad v > |u - x|\}.$$

Note that this condition involves only the modulus of the coefficients $\alpha(\zeta)$.

If the integral $\int_{-\infty}^{\infty} (A(x))^p dx$ is finite, then necessarily

$$f(z) = \iint_P (z - \bar{\zeta})^{-2} \alpha(\zeta) du\, dv \quad \text{belongs to} \quad H^p(\mathbb{R}),$$

and if $1 \le p < \infty$,

$$(2.28) \qquad \|f\|_p \le C(p) \left(\int_{-\infty}^{\infty} (A(x))^p dx \right)^{1/p}.$$

The left member of (2.28) is the norm of f in $H^p(\mathbb{R})$, as defined by (2.25). This estimate, however, is sometimes very crude. If, for example, $f(z) = (z + i)^{-2}$, one is led to choose the Dirac measure at the point i for $\alpha(\zeta)$, and the second member of (2.28) is infinite. This paradox arises because the representation

$$(2.29) \qquad f(z) = \iint_P (z - \bar{\zeta})^{-2} \alpha(\zeta) du\, dv$$

is not unique.

To obtain a unique decomposition, which we call the natural decomposition, we restrict the choice to $\alpha(\zeta) = \frac{2i}{\pi} v f'(u + iv)$. When we do this, the two norms $\|f\|_p$ and $\|A\|_p$ become equivalent if $1 \le p < \infty$. Today this choice of natural coefficients has an interesting explanation. This interpretation, which depends on the contemporary formalism of wavelet theory, is given in the following section.

2.7.   Atomic decompositions, from 1960 to 1980.

Guido Weiss, in collaboration with Ronald R. Coifman, was the first to interpret, as we have just done, Lusin's theory in terms of "atoms" and "atomic decompositions." The "atoms" are the "simplest elements" of a function space, and the objective of the theory is to find, for the usual function spaces, (1) the atoms and (2) the "assembly rules" that allow the reconstruction of all the elements of the function space using these atoms.

In the case of the holomorphic Hardy spaces of the last section, the atoms were the functions $(z - \bar{\zeta})^{-2}$, $\zeta \in P$, and the assembly rules were given by the condition on Lusin's area function $A(x)$.

For the spaces $L^p[0, 2\pi]$, $1 < p < \infty$, the "atoms" cannot be the functions $\cos kx$ and $\sin kx$, $k \ge 1$, because this choice does not lead to assembly rules that are sufficiently simple and explicit to be useful in practice. Marcinkiewicz showed in 1938 that the simplest atomic decomposition for the spaces $L^p[0, 1]$, $1 < p < \infty$, is given by the Haar system. The Franklin basis would have served as well and, from the scientific perspective given by wavelet theory, the Franklin basis and Littlewood–Paley analysis are naturally related.

One of the approaches to "atomic decompositions" is given by Calderón's identity. To explain Calderón's identity, we start with a function $\psi(x)$ belonging to $L^2(\mathbb{R}^n)$. Later in this history, Grossmann and Morlet called this function an analyzing wavelet. Its Fourier transform $\hat{\psi}(\xi)$ is subject to the condition that

$$(2.30) \qquad \int_0^{\infty} |\hat{\psi}(t\xi)|^2 \frac{dt}{t} = 1 \quad \text{for almost all} \quad \xi \in \mathbb{R}^n.$$

If $\psi(x)$ belongs to $L^1(\mathbb{R}^n)$, condition (2.30) implies

$$(2.31) \qquad\qquad \int_{\mathbb{R}^n} \psi(x)dx = 0.$$

We write $\tilde{\psi}(x) = \overline{\psi(-x)}$, $\psi_t(x) = t^{-n}\psi(x/t)$, and $\tilde{\psi}_t(x) = t^{-n}\tilde{\psi}(x/t)$. Let Q_t denote the operator defined as convolution with ψ_t; its adjoint Q_t^* is the operator defined as convolution with $\tilde{\psi}_t$.

Calderón's identity is a decomposition of the identity operator, written as

$$(2.32) \qquad\qquad I = \int_0^\infty Q_t Q_t^* \frac{dt}{t}.$$

Grossmann and Morlet rediscovered this identity in 1980, 20 years after the work of Calderón. However, with this rediscovery, they gave it a different interpretation by relating it to "the coherent states of quantum mechanics" [6].

They defined wavelets (generated from the analyzing wavelet ψ) by

$$(2.33) \qquad \psi_{(a,b)}(x) = a^{-n/2}\psi\left(\frac{x-b}{a}\right), \qquad a > 0, \qquad b \in \mathbb{R}^n.$$

In the analysis and synthesis of an arbitrary function $f(x)$ belonging to $L^2(\mathbb{R}^n)$, these wavelets $\psi_{(a,b)}$ are going to play the role of an orthonormal basis. The wavelet coefficients $W(a,b)$ are defined by

$$(2.34) \qquad\qquad W(a,b) = \langle f, \psi_{(a,b)} \rangle,$$

where $\langle u, v \rangle = \int u(x)\overline{v(x)}\,dx$.

The function $f(x)$ is *analyzed* by (2.34). The *synthesis* of $f(x)$ is given by

$$(2.35) \qquad f(x) = \int_0^\infty \int_{\mathbb{R}^n} W(a,b)\psi_{(a,b)}(x)db\frac{da}{a^{n+1}}.$$

This is a linear combination of the original wavelets using the coefficients given by the analysis.

We return to the specific case of the Hardy spaces $H^p(\mathbb{R})$ for $1 \le p < \infty$. The analyzing wavelet $\psi(x) = \frac{1}{\pi}(x+i)^{-2}$ chosen by Lusin is the restriction to the real axis of the function $\frac{1}{\pi}(z+i)^{-2}$; it is holomorphic in P and belongs to all of the Hardy spaces. The Fourier transform of ψ is $\hat{\psi}(\xi) = -2\xi e^{-\xi}$ for $\xi \ge 0$ and $\hat{\psi}(\xi) = 0$ if $\xi \le 0$.

Condition (2.30) is not satisfied but, on the other hand, we have

$$(2.36) \qquad \int_0^\infty |\hat{\psi}(t\xi)|^2 \frac{dt}{t} = 1 \quad \text{if} \ \ \xi > 0, \quad = 0 \ \ \text{if} \ \ \xi \le 0.$$

Condition (2.36) implies that the wavelets $\psi_{(a,b)}$ generate $H^2(\mathbb{R})$ instead of $L^2(\mathbb{R})$ when $a > 0$, $b \in \mathbb{R}$.

The wavelet coefficients of a function $f(x)$ belonging to the Hardy space $H^2(\mathbb{R})$ are then

$$(2.37) \qquad W(a,b) = \langle f, \psi_{(a,b)} \rangle = \frac{1}{\pi} \int_{-\infty}^{\infty} f(x) \frac{a\sqrt{a}}{(x-b-ia)^2} \, dx,$$

which is equal to $2ia\sqrt{a}f'(b+ia)$ since $f(z)$ is holomorphic in P.

Thus the representation (2.35) of a function in the Hardy space $H^2(\mathbb{R})$ coincides with the "natural representation" that we defined in the preceding section.

Note that $\psi(x) = \frac{1}{x+i}$, although it belongs to all the Hardy spaces, cannot be used as an analyzing wavelet because

$$\int_0^{\infty} |\hat{\psi}(t\xi)|^2 \frac{dt}{t} = +\infty \quad \text{if} \quad \xi > 0.$$

2.8. Strömberg's wavelets.

The real version of the holomorphic Hardy space $H^1(\mathbb{R})$ is denoted by $\mathcal{H}^1(\mathbb{R})$ and is composed of real-valued functions $u(x)$ such that $u(x) + iv(x)$ belongs to $H^1(\mathbb{R})$, where $v(x)$ is also real-valued. In other words, $u(x)$ belongs to $\mathcal{H}^1(\mathbb{R})$ if and only if u and its Hilbert transform $\tilde{u}$ belong to $L^1(\mathbb{R})$.

Research on "atomic decompositions" for the functions in the Hardy space $\mathcal{H}^1(\mathbb{R})$ takes two completely different approaches: one involves the atomic decomposition of Coifman and Weiss, and the other concerns the search for unconditional bases for the space $\mathcal{H}^1$. Here is an outline of these theories.

Coifman and Weiss showed that the most general function f of $\mathcal{H}^1$ can be written as $f(x) = \sum_0^{\infty} \lambda_k a_k(x)$, where the coefficients λ_k are such that $\sum_0^{\infty} |\lambda_k| < \infty$ and where the $a_k(x)$ are atoms of $\mathcal{H}^1$. This means that for each $a_k(x)$, there exists an interval I_k such that $a_k(x) = 0$ outside of I_k, $|a_k(x)| \leq 1/|I_k|$ ($|I_k|$ is the length of I_k), and $\int_{I_k} a_k(x)dx = 0$. These three conditions imply that the norms of the a_k in $\mathcal{H}^1$ are bounded by a fixed constant C_0. The price to pay for this extraordinary decomposition is that it is not given by a linear algorithm.

Finding an unconditional basis means constructing, once and for all, a sequence of functions $b_k(x)$ of $\mathcal{H}^1$ that are linearly independent, in a very strong sense, and that allow all functions f of $\mathcal{H}^1$ to be decomposed as

$$(2.38) \qquad f(x) = \sum_0^{\infty} \beta_k b_k(x),$$

where the scalars β_k are defined explicitly by the formulas

$$(2.39) \qquad \beta_k = \int f(x) g_k(x) dx.$$

Here the g_k are specific functions in the dual of $\mathcal{H}^1$, which is to say they are specific BMO functions.

The strong independence property is this: There exists a constant C such that if two sequences of coefficients β_k and λ_k verify $|\beta_k| \leq |\lambda_k|$ for all k, then

$$(2.40) \qquad \left\| \sum_0^\infty \beta_k b_k(x) \right\| \leq C \left\| \sum_0^\infty \lambda_k b_k(x) \right\|,$$

where $\| \cdot \|$ is the norm of the function space $\mathcal{H}^1$.

Wojtasczyk proved that the Franklin system $f_0(x), f_1(x), \ldots, f_n(x), \ldots$, without the function 1, is an unconditional basis for the subspace of $\mathcal{H}^1(\mathbb{R})$ composed of functions that vanish outside the interval $[0,1]$.

Strömberg showed that the orthonormal wavelet basis defined by (2.24) is, in fact, an unconditional basis for the space $\mathcal{H}^1(\mathbb{R})$.

Does there exist a relation between these two types of atomic decompositions? To construct the decomposition of Coifman and Weiss using an unconditional basis, it is necessary, first of all, to use a wavelet with compact support in place of the one used by Strömberg.

The discovery of orthonormal bases of the form $2^{j/2}\psi(2^j x - k)$, $j, k \in \mathbb{Z}$, where $\psi(x)$ is in the class C^1 and has compact support, is due to Ingrid Daubechies and dates from 1987. This will be developed a little later in §2.9.

To obtain the Coifman and Weiss decomposition of a function f in $\mathcal{H}^1$, we first decompose it in Daubechies's orthonormal basis, $\psi_{j,k}$. The series obtained is regrouped into blocks. These blocks are a little like the dyadic blocks of Littlewood and Paley; however, this time, they are defined by considering the moduli of the coefficients $\alpha_{j,k}$ of this series. The interested reader is referred to [8].

2.9. A first synthesis: Wavelet analysis.

Thanks to the historical perspective that we have today, we can relate the Littlewood–Paley decomposition (1930), the version of Franklin's basis given by Strömberg (1927), and Calderón's identity (1960).

This first synthesis will be followed by a more inclusive synthesis that encompasses the techniques of numerical signal and image processing. This second synthesis will lead to Daubechies's orthonormal bases.

This first synthesis is based on the definition of the word "wavelet" and on the concept of "wavelet transform." We will see that the success of this synthesis depends on a certain lack of specificity in the original definition. At this time, mathematicians had not created a general formalism covering all of the examples we presented above. A physicist and an engineer, Grossmann and Morlet, provided a definition and a way of thinking based on physical intuition that was flexible enough to cover all these cases. Starting with the Grossmann–Morlet definition, we will present two other definitions and indicate how they are related.

The first definition of a wavelet, which is due to Grossmann and Morlet, is quite broad. *A wavelet is a function ψ in $L^2(\mathbb{R})$ whose Fourier transform $\hat{\psi}(\xi)$ satisfies the condition $\int_0^\infty |\hat{\psi}(t\xi)|^2 \frac{dt}{t} = 1$ almost everywhere.*

The second definition of a wavelet is adapted to the Littlewood–Paley–Stein theory. A *wavelet is a function* ψ *in* $L^2(\mathbb{R}^n)$ *whose Fourier transform* $\hat{\psi}(\xi)$ *satisfies the condition* $\sum_{-\infty}^{\infty} |\hat{\psi}(2^{-j}\xi)|^2 = 1$ *almost everywhere.* If ψ is a wavelet in this sense, then $\sqrt{\log 2}\,\psi$ satisfies the Grossmann–Morlet condition.

The third definition refers to the work of Franklin and Strömberg. A *wavelet is a function* ψ *in* $L^2(\mathbb{R})$ *such that* $2^{j/2}\psi(2^j x - k)$, $j, k \in \mathbb{Z}$, *is an orthonormal basis for* $L^2(\mathbb{R})$. Such a wavelet ψ necessarily verifies the second condition.

This shows that in going from the first to the third definition we are adding more conditions and thus narrowing the scope of "wavelet." The same is true for the wavelet analysis of a function. In the general Grossmann–Morlet theory (which is identical to Calderón's theory) the wavelet analysis of a function f yields a function $W(a, b)$ of $n + 1$ variables $a > 0$ and $b \in \mathbb{R}^n$. This function is defined by (2.34): $W(a, b) = \langle f, \psi_{(a,b)} \rangle$, where $\psi_{(a,b)}(x) = a^{-n/2}\psi(\frac{x-b}{a})$, $a > 0$, $b \in \mathbb{R}^n$.

In the Littlewood–Paley theory, a is replaced by 2^{-j}, while b is denoted by x. In other words, if Γ is the multiplicative group $\{2^{-j}, j \in \mathbb{Z}\}$, then the Littlewood–Paley analysis is obtained by restricting the Grossmann–Morlet analysis to $\Gamma \times \mathbb{R}^n$.

In the Franklin–Strömberg theory, a is replaced by 2^{-j} and b is replaced by ka^{-j}, where $j, k, \in \mathbb{Z}$. In other words, the analysis of f in the Franklin–Strömberg basis is obtained by restricting the Littlewood–Paley analysis to the "hyperbolic lattice" S in $(0, \infty) \times \mathbb{R}$ consisting of the points $(2^{-j}, k2^{-j})$, $j, k \in \mathbb{Z}$. The logical relations among these wavelets analyses are easy to verify.

We start with the Grossmann–Morlet analysis, that is, the Calderón identity. This is written $I = \int_0^\infty Q_t Q_t^* \frac{dt}{t}$, where $Q_t(f) = f * \psi_t$. This becomes $I = \sum_{-\infty}^{\infty} \Delta_j \Delta_j^*$ in the Littlewood–Paley theory, and if $t = 2^{-j}$, one has $Q_t(f) = \Delta_j(f)$. Replacing t by 2^{-j} and the integral $\int_0^\infty u(t)\frac{dt}{t}$ by the sum $\sum_{-\infty}^{\infty} u(2^{-j})$ is completely classic.

To relate Littlewood–Paley analysis to the analysis that is obtained using the orthogonal wavelets of Franklin and Strömberg, we write $\psi_j(x) = 2^j \psi(2^j x)$, and $\tilde{\psi}_j(x) = 2^j \tilde{\psi}(2^j x)$, where $\tilde{\psi}(x) = \overline{\psi(-x)}$. We let $\Delta_j(f)$ denote the convolution product $f * \psi_j$ and $\Delta_j^* : L^2(\mathbb{R}) \to L^2(\mathbb{R})$ the adjoint of the operator $\Delta_j : L^2(\mathbb{R}) \to L^2(\mathbb{R})$. Then $\Delta_j^*(f) = f * \tilde{\psi}_j$. The coefficients $\alpha(j, k)$ of the decomposition of f in Strömberg's orthonormal basis are then given by

(2.41)
$$\alpha(j, k) = 2^{j/2} \int f(x)\overline{\psi(2^j x - k)}\, dx$$
$$= 2^{-j/2}(f * \tilde{\psi}_j)(k2^{-j}) = 2^{-j/2}\Delta_j^* f(k2^{-j}).$$

Thus the coefficients are obtained by restricting $\Delta_j^*(f)$ to the hyperbolic lattice S.

In all three cases, wavelet analysis is followed by a synthesis that reconstructs $f(x)$ from its wavelet transform. In the case of Grossmann–Morlet wavelets, this synthesis is given by the identity (2.35), which we rewrite here:

$$(2.42) \qquad f(x) = \int_0^\infty \int_{\mathbb{R}^n} W(a,b)\psi_{(a,b)}(x)db\frac{da}{a^{n+1}}.$$

In the case of the Littlewood–Paley analysis, the integral $\int_0^\infty u(a)\frac{da}{a}$ is replaced, as we have already mentioned, by the sum $\sum_{-\infty}^{\infty} u(2^{-j})$ and (2.42) becomes

$$(2.43) \qquad f(x) = \sum_{-\infty}^{\infty} \int_{\mathbb{R}^n} (\Delta_j^* f)(b)\psi_j(x-b)db.$$

Finally, for the one-dimensional case and for Strömberg's orthogonal wavelets, the last integral becomes a sum, and (2.43) becomes

$$(2.44) \qquad f(x) = \sum_{-\infty}^{\infty}\sum_{-\infty}^{\infty} \alpha(j,k)\psi_{j,k}(x).$$

All of the preceding arguments remain at a fairly superficial level since the hypothesis on ψ enables us to analyze only the space L^2 of square-summable functions. This is the setting in which Grossmann and Morlet wrote their theoretical work. But this is evidently a sort of regression, for we have just shown that, across a century of mathematical history, wavelet analysis was created specifically to analyze function spaces other than L^2. Fourier analysis serves admirably for L^2.

If we want wavelets to be useful for the analysis of these other function spaces, it is necessary to impose conditions on them in addition to those we have already given. Up until now we have required only that the analysis preserve energy or, which is the same thing, that the synthesis give an exact reconstruction.

These new conditions are

$$(2.45) \qquad \text{the regularity of the wavelet } \psi,$$
$$(2.46) \qquad \text{the decay at infinity of } \psi, \text{ and}$$
$$(2.47) \qquad \text{the number of vanishing moments.}$$

We can, for example, impose the conditions that ψ belongs to the Schwartz class and that all of its moments vanish. We can also require (which will be the case for Daubechies's wavelets) that $\psi(x)$ have m continuous derivatives, that it have compact support, and that its first $r+1$ moments vanish.

The properties of the Strömberg wavelet are intermediate: It has exponential decay, as does its first derivative, and

$$\int_{-\infty}^{\infty} \psi(x)dx = \int_{-\infty}^{\infty} x\psi(x)dx = 0.$$

2.10.   The advent of signal processing.

If history had stopped with this first synthesis, the Daubechies orthonormal bases, which improve the rudimentary Haar basis, would never have been discovered.

A new start was made in 1985 by Stephane Mallat, a specialist in numerical image processing. Mallat discovered some close relations among

(a) the quadrature mirror filters, which were invented by Croissier, Esteban, and Galand for the digital telephone.

(b) the pyramid algorithms of Burt and Adelson, which are used in the context of numerical image processing, and

(c) the orthonormal wavelet bases discovered by Strömberg and his successors.

These relations will be explained in the next two chapters. Using "Mallat's program," Daubechies was able to complete Haar's work. For each integer r, Daubechies constructs an orthonormal basis for $L^2(\mathbb{R})$ of the form

$$(2.48) \qquad\qquad 2^{j/2}\psi_r(2^j x - k), \qquad j, k \in \mathbb{Z},$$

having the following properties:

$$(2.49) \qquad\qquad \text{the support of } \psi_r \text{ is the interval } [0, 2r+1],$$

$$(2.50) \qquad 0 = \int_{-\infty}^{\infty} \psi_r(x)\,dx = \cdots = \int_{-\infty}^{\infty} x^r \psi_r(x)\,dx, \quad \text{and}$$

$$(2.51) \qquad\qquad \psi_r(x) \text{ has } \gamma r \text{ continuous derivatives,}$$

where the positive constant γ is about $1/5$. When $r = 0$, this reduces to the Haar system.

Daubechies's wavelets provide a much more effective analysis and synthesis than that obtained with the Haar system. If the function being analyzed has m continuous derivatives, where $0 \le m \le r + 1$, then the coefficients $\alpha(j, k)$ from its decomposition in the Daubechies basis will be of the order of magnitude $2^{-(m+1/2)j}$, while it would be of the order $2^{-3j/2}$ with the Haar system. This means that as soon as the analyzed function is regular, the coefficients one keeps (those exceeding the machine precision) will be much less numerous than in the case of the Haar system. Thus one speaks of signal "compression." Furthermore, this property has a purely local aspect because Daubechies's wavelets have compact support.

Synthesis using Daubechies's wavelets also gives better results than the Haar system. In the latter case, a regular function is approximated by functions that have strong discontinuities. This is very annoying for image processing, as the reader can verify by referring to the image of Abraham Lincoln on page 74 of Marr's book [7]. These remarkable qualities of Daubechies's bases explain their undisputed success.

2.11. Conclusions.

The status of "wavelet analysis" within mathematics is rather unique. Indeed, mathematicians have been working on wavelets, which were called "atomic decompositions," for a fairly long time. Their goal was to provide direct and easy access to the various function spaces. But during all of this period, which stretches from 1909 to 1980, from Haar to Strömberg, there was very little scientific interchange among mathematicians (of the "Chicago School"), physicists, and experts in signal processing. Not knowing about the mathematical developments and faced with the pressure of specific needs within their disciplines, the last two groups were led to rediscover wavelets.

For example, Marr did not know about Calderón's work on wavelets (dating from 1960) when he announced the hypothesis that we analyze in detail in Chapter 8. Similarly, G. Battle and P. Federbush [2] were not aware of Strömberg's basis when they needed it to do renormalization computations in quantum field theory.

In the numerous fields of science and technology where wavelets appeared at the end of the 1970s, they were handcrafted by the scientists and engineers themselves. Their use has never resulted from proselytism by mathematicians.

Today the boundaries between mathematics and signal and image processing have faded, and mathematics has benefited from the rediscovery of wavelets by experts from other disciplines. The detour through signal and image processing was the most direct path leading from the Haar basis to Daubechies's wavelets.

Bibliography

[1] G. BATTLE, *A block spin construction of ondelettes, Part II: The QFT connection*, Comm. Math. Phys., 114 (1988), pp. 93–102.

[2] G. BATTLE AND P. FEDERBUSH, *Ondelettes and phase cluster expansions, a vindication*, Comm. Math. Phys., 109 (1987), pp. 417–419.

[3] A. BENASSI, S. JAFFARD, AND D. ROUX, *Analyse multi-échelle des champs gaussiens markoviens d'ordre p indexés par* [0,1], C.R. Acad. Sci. Paris, Sér. I (1991), pp. 403–406.

[4] C. FEFFERMAN, *The multiplier problem for the ball*, Ann. of Math., 94 (1971), pp. 330–336.

[5] P. FLANDRIN, *Wavelet analysis and synthesis of fractional Brownian motion*, preprint, Ecole Normale Supérieure de Lyon, Lab. de Physique (URA 1325 CNRS), Lyon, France, 1991.

[6] A. GROSSMANN AND J. MORLET, *Decomposition of Hardy functions into square integrable wavelets of constant shape*, SIAM J. Math., 15 (1984), pp. 723–736.

[7] D. MARR, *Vision*, Freeman and Co., New York, 1982.

[8] Y. MEYER, *Ondelettes*, Hermann, New York, 1990.

[9] A. PAPOULIS, *Signal Analysis*, 4th edition, McGraw-Hill, New York, 1988.

[10] J. VILLE, *Théorie et applications de la notion de signal analytique*, C&T, Laboratoire de Télécommunications de la Société Alsacienne de Construction Mécanique, 2éme A. No. 1, (1948).

CHAPTER 3

Quadrature Mirror Filters

3.1. Introduction.

It was with keen pleasure that I recently reread Claude Galand's thesis entitled "Codage en sous-bandes: théorie et applications à la compression numérique du signal de parole." This thesis was defended on 25 March 1983 at the Signals and Systems Laboratory of the University of Nice.

In his thesis, Galand carefully describes the quadrature mirror filters (which he invented in collaboration with Esteban and Croissier) and their anticipated applications. He also posed some very important problems that would lead to the discovery of "wavelet packets" (Chapter 7) and "Malvar's wavelets" (Chapter 6).

Galand's work was motivated by the possibility of improving the digital telephone, a technology that involves transmitting speech signals as sequences of 0's and 1's. However, as Galand remarked, these techniques extend far beyond the digital telephone; facsimile, video, databases, etc., all travel over telephone lines.

At present, the bit allocation used for telephone transmission is the well-known 64 kilobits per second. Galand sought, by using coding methods tailored to speech signals, to transmit speech well below this standard.

To validate the method he proposed, Galand compared it to two other techniques for coding sampled speech: predictive coding and transform coding.

Linear prediction coding amounts to looking for the correlations among successive values of the sampled signal. These correlations are likely to occur on intervals of the order of 20 to 30 milliseconds. This leads one to cut the sampled signal $x(n)$ into blocks defined by $1 \leq n \leq N$, $N+1 \leq n \leq 2N$, etc., and then to seek, for each block, coefficients a_k, $1 \leq k \leq p$, that minimize the quadratic mean $\frac{1}{N} \sum_1^N |e(n)|^2$ of the prediction errors defined by

$$e(n) = x(n) - \sum_{k=1}^p a_k x(n-k).$$

In general, p is much smaller than N. To transmit the block $x(n)$, $1 \leq n \leq N$, it suffices to transmit the first p values $x(1), \ldots, x(p)$, the p coefficients $a_1, \ldots, a_p$, and the prediction errors $e(n)$. The method is efficient if most of the prediction errors are near 0. When they fall below a certain threshold, they are not transmitted, and significant compression can result.

Transform coding consists in cutting the sampled signal into successive blocks of length N, as we have just done, and then using a unitary transformation A to transform each block (denoted by X) into another block (denoted by Y). The block Y is then quantized, with the hope that, for a suitable linear transformation, the Y blocks will have a simpler structure than the X blocks. *Subband coding* will be presented in the next section.

For a stationary Gaussian signal, the theoretical limits of the minimal distortion that can be obtained by the three methods are the same. But, as Galand showed, this assumes, in the case of subband coding, that the width of the frequency channels tends to 0 and that their number tends to ∞. We will show that these conditions cannot be satisfied (rigorously) for subband coding. To provide an approximate solution, Galand proposed a treelike arrangement of quadrature mirror filters. This construction leads precisely to the wavelet packets, which we discuss in detail in Chapter 7.

In the cases of linear prediction coding and transform coding, the theoretical limits of the minimal distortion are calculated as the lengths of the blocks tend to infinity, while conserving the stationarity hypothesis.

If the three types of coding yield asymptotically the same quality of compression, why introduce subband coding? Galand saw two advantages: the simplicity of the algorithm and the possibility that subband coding would reduce the unpleasant effects of quantization noise as perceived at the receiver. By quantizing inside each subband, the signal would tend to mask the quantization noise, and it would be less apparent.

The same argument has been repeated by Adelson, Hingorani, and Simoncelli [1] for numerical image processing. The use of pyramid algorithms and wavelets allows aspects of the human visual system to be cleverly taken into account so that the signal masks the noise. The perceptual quality of the reconstructed image is improved even though the theoretical compression calculations do not distinguish this method from the others.

3.2. Subband coding: The case of ideal filters.

We follow Galand's example and begin with a deceptively simple case. For a fixed $m \geq 2$, let I denote an interval of length $2\pi/m$ within $[0, 2\pi]$, and let l_I^2 denote the Hilbert space of sequence $(c_k)_{k \in \mathbb{Z}}$ verifying $\sum |c_k|^2 < \infty$ and such that $f(\theta) = \sum_{-\infty}^{\infty} c_k e^{ik\theta}$ is zero outside the interval I. This subspace l_I^2 will be called a frequency channel.

If $(c_k)_{k \in \mathbb{Z}}$ is a sequence belonging to l_I^2, the subsequence $(c_{mk})_{k \in \mathbb{Z}}$ provides an optimal, compact representation of the original because

$$\frac{1}{m} f(\theta) = \frac{1}{m} \left(f(\theta) + f\left(\theta + \frac{2\pi}{m}\right) + \cdots + f\left(\theta + \frac{2\pi(m-1)}{m}\right) \right) = \sum c_{km} e^{im\theta}$$

if $f(\theta)$ vanishes outside I and if $\theta \in I$.

Thus we have $\sum_{-\infty}^{\infty} |c_{km}|^2 = \frac{1}{m} \sum_{-\infty}^{\infty} |c_k|^2$, and this relation expresses the redundancy contained in the original sequence (c_k), which is strongly correlated.

This means that the original sequence contains m times the numerical data needed to reconstruct $f(\theta)$ on I, knowing that $f(\theta)$ is 0 outside of I.

The *ideal scheme* for subband coding consists in first *filtering* the incoming signal into m frequency channels associated with the intervals $[2\pi l/m, 2\pi(l+1)/m]$, $0 \leq l \leq m-1$, and then *subsampling* the corresponding outputs, retaining only one point in m. This operation, which consists in restricting a sequence defined on $\mathbb{Z}$ to $m\mathbb{Z}$, is called *decimation* and is denoted by $m \downarrow 1$. Thus the ideal subband coding scheme is illustrated as follows:

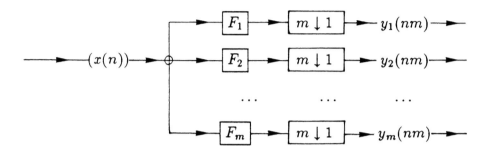

The scheme for *reconstructing* the original signal is the dual of the analysis scheme. We began by extending the sequences $y_1(nm), \ldots, y_m(nm)$ by inserting 0's at all integers that are not multiples of m. Next we filter this "absurd decision" by again using the filters $F_1, \ldots, F_m$. The output returns $(x(n))$.

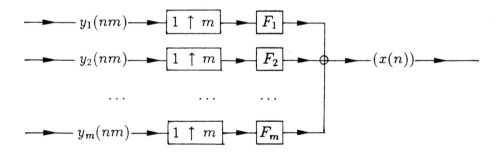

One can, as Galand did, hope for the best, and try to replace the index functions of the intervals $[2\pi l/m, 2\pi(l+1)/m]$ with more regular functions of the form $w(mx - 2\pi l)$, $0 \leq 1 \leq m-1$. If the $w(x) = w_m(x)$ were a finite trigonometric sum, then the filters $F_1, \ldots, F_m$ would have finite length, which is essential for applications.

But the Balian–Low theorem (Chapter 6) tells us that such a function $w(x)$ cannot be constructed if we demand that it be regular and well localized (uniformly in m). Consequently, it is not possible to realize the ideal subband coding scheme just described if we require that the filters $F_1, \ldots, F_m$ have finite length and, at the same time, provide good frequency definition.

3.3.   Quadrature mirror filters.

Faced with the impossibility of realizing subband coding using m bands covering the frequency space regularly and having finite-length filters (whose length must be Cm, as required by the Heisenberg uncertainty principle), Galand limited himself to the case $m = 2$. He then had the idea to effect a finer frequency tiling by suitably iterating the two-band process. We will see in Chapter 7 that this arborescent scheme leads directly to "wavelet packets," but we will also see that these "wavelet packets" do not have the desired spectral properties.

Subband coding using two frequency channels works perfectly. We are going to describe it in detail.

The input signals $x(n)$ have already been sampled and are defined on the integers $n \in \mathbb{Z}$; they are arbitrary sequences with finite energy:

$$\sum_{-\infty}^{\infty} |x(n)|^2 < \infty.$$

We denote by $D : l^2(\mathbb{Z}) \to l^2(2\mathbb{Z})$ the decimation operator (also denoted by $2 \downarrow 1$), which consists in retaining only the terms with even index in the sequence $(x(n))_{n \in \mathbb{Z}}$. The adjoint operator

$$E = D^* : l^2(2\mathbb{Z}) \to l^2(\mathbb{Z})$$

is the crudest possible extension operator. It consists, starting with a sequence $(x(2n))_{n \in \mathbb{Z}}$, in constructing the sequence defined on $\mathbb{Z}$ obtained by inserting 0's at the odd indices. Thus we get the sequence

$$(\ldots, x(-2), 0, x(0), 0, x(2), 0, x(4), 0, \ldots).$$

To simplify the notation we write X in place of $(x(n))_{n \in \mathbb{Z}}$. These input signals X are first filtered using two filters F_0 and F_1.

We will see later that F_0 must be a low-pass filter (in a sense that will be made precise) and that F_1 will then be a high-pass filter.

The outputs $X_0 = F_0(X)$ and $X_1 = F_1(X)$ are two signals $(x_0(n))_{n \in \mathbb{Z}}$ and $(x_1(n))_{n \in \mathbb{Z}}$ with finite energy.

X_0 and X_1 are subsampled with the decimation operator $D = 2 \downarrow 1$. Then we have $Y_0 = D(X_0) = (x_0(2n))_{n \in \mathbb{Z}}$ and $Y_1 = D(X_1) = (x_1(2n))_{n \in \mathbb{Z}}$.

Write

$$\|Y_0\| = \left(\sum_{-\infty}^{\infty} |x_0(2n)|^2 \right)^{1/2} \quad \text{and} \quad \|Y_1\| = \left(\sum_{-\infty}^{\infty} |x_1(2n)|^2 \right)^{1/2}.$$

The two filters F_0 and F_1 are called quadrature mirror filters if, for all signals X of finite energy, one has

(3.1) $$\|Y_0\|^2 + \|Y_1\|^2 = \|X\|^2.$$

Denote by T_0 the operator $DF_0 : l^2(\mathbb{Z}) \to l^2(2\mathbb{Z})$ and, similarly, let T_1 denote the operator $DF_1 : l^2(\mathbb{Z}) \to l^2(2\mathbb{Z})$.

Then (3.1) is clearly equivalent to

(3.2) $$I = T_0^* T_0 + T_1^* T_1.$$

In (3.2), $I : l^2(\mathbb{Z}) \to l^2(\mathbb{Z})$ is the identity operator. What is much less evident is that the vectors $T_0^* T_0(X)$ and $T_1^* T_1(X)$ are always orthogonal, as the following theorem asserts.

THEOREM 3.1. Let $F_0(\theta)$ and $F_1(\theta)$ denote the transfer functions of the filters F_0 and F_1. Then the following two properties are equivalent to each other and to (3.1).

(3.3) The matrix $\dfrac{1}{\sqrt{2}} \begin{pmatrix} F_0(\theta) & F_1(\theta) \\ F_0(\theta + \pi) & F_1(\theta + \pi) \end{pmatrix}$
is unitary for almost all $\theta \in [0, 2\pi]$.

(3.4) The operator $(T_0, T_1) : l^2(\mathbb{Z}) \to l^2(2\mathbb{Z}) \times l^2(2\mathbb{Z})$
is an isometric isomorphism.

Recall that the sequence of Fourier coefficients of the 2π-periodic function $F_0(\theta)$ is the impulse response of the filter F_0, and the same is true for $F_1(\theta)$ and F_1.

Here are some comments on these different properties.

Condition (3.2) is called the perfect reconstruction property: The input signal X is the sum of two orthogonal signals $T_0^* T_0(X)$ and $T_1^* T_1(X)$, where the signals $T_0(X)$ and $T_1(X)$ were given by the analysis. The operators $T_0^* = F_0^* E$ and $T_1^* = F_1^* E$ are applied to two sequences sampled on the even integers. These are first extended in the crudest way, which is by replacing the missing values with 0's. Next, this seemingly nonsensical step is corrected by passing the sequences through the filters F_0^* and F_1^*, which are the adjoints of F_0 and F_1. The correct result is read at the output. The complete scheme, analysis and synthesis, is now classic:

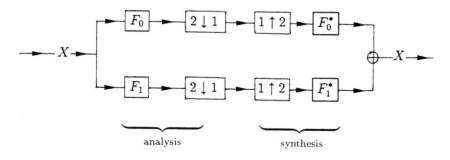

Condition (3.4) means that quadrature mirror filters constitute orthogonal transformations of a particular type, while (3.3) allows us to construct quadrature mirror filters that have finite impulse response.

To see this, we start with a trigonometric polynomial $m_0(\theta) = \alpha_0 + \alpha_1 e^{i\theta} + \cdots + \alpha_N e^{iN\theta}$ such that $|m_0(\theta)|^2 + |m_0(\theta+\pi)|^2 = 1$ for all θ. (We will see further on how to construct these polynomials.) Next, we write $F_0(\theta) = \sqrt{2}\,\overline{m_0(\theta)}$ and $F_1(\theta) = \sqrt{2}\,e^{i\theta}m_0(\theta + \pi)$. Then it follows directly that (3.3) is satisfied.

The following *five examples* illustrate the definition of quadrature mirror filters.

The *first example* is essentially a counterexample because it is never used, for a reason that will become clear later. It consists in bypassing the operators F_0 and F_1. More precisely, $T_0(X)$ is the restriction of the sequence $X = (x(n))_{n \in \mathbb{Z}}$ to the even integers, while $T_1(X)$ is the restriction of this sequence to the odd integers. Condition (3.1) is then trivially satisfied, and the unitary matrix (3.3) is

$$\frac{1}{\sqrt{2}} \begin{pmatrix} 1 & e^{i\theta} \\ 1 & -e^{i\theta} \end{pmatrix}.$$

The *second example* is more interesting. The filter operators are defined by

$$F_0((x(n))) = \frac{1}{\sqrt{2}}(x(n) + x(n+1)) \quad \text{and}$$

$$F_1((x(n))) = \frac{1}{\sqrt{2}}(x(n) - x(n+1)).$$

The orthonormal wavelet basis associated with this choice will be the Haar system.

The *third example* recaptures the ideal filters presented at the beginning of the chapter. The 2π-periodic function $m_0(\theta)$ is 1 on $[0,\pi)$ and 0 on $[\pi, 2\pi)$, and $m_1(\theta) = 1 - m_0(\theta)$. Next, as above, define $F_0(\theta) = \sqrt{2}\,\overline{m_0(\theta)}$ and $F_1(\theta) = \sqrt{2}\,\overline{m_1(\theta)}$.

In the *fourth example*, $m_0(\theta)$ becomes the characteristic function of $[-\pi/2, \pi/2)$ when it is restricted to $[-\pi, \pi)$, and $m_1(\theta) = 1 - m_0(\theta)$.

The *last example* is a smooth modification of the preceding one. Assume that $0 < \alpha < \pi/2$. We ask that $m_0(\theta)$ be 2π-periodic, equal to 1 on the interval $[-\frac{\pi}{2} + \alpha, \frac{\pi}{2} - \alpha]$, equal to 0 on $[\frac{\pi}{2} + \alpha, \frac{3\pi}{2} - \alpha]$, even, and infinitely differentiable. In addition, we impose the condition

(3.5) $|m_0(\theta)|^2 + |m_0(\theta + \pi)|^2 = 1.$

Then write $m_1(\theta) = e^{-i\theta}\overline{m_0(\theta + \pi)}$, $F_0(\theta) = \sqrt{2}\,\overline{m_0(\theta)}$, $F_1(\theta) = \sqrt{2}\,\overline{m_1(\theta)}$, and we obtain two quadrature mirror filters.

3.4.   The trend and fluctuation.

Returning to Theorem 3.1, let H denote the Hilbert space of all sequences $(x(n))_{n \in \mathbb{Z}}$ satisfying $\sum_{-\infty}^{\infty} |x(n)|^2 < \infty$, and recall the operators T_0 and T_1. Write H_0 and H_1 for the two subspaces $T_0^* T_0(H)$ and $T_1^* T_1(H)$. We know that if F_0 and F_1 are two quadrature mirror filters, then H will be the direct orthogonal sum of H_0 and H_1.

Write $m_0(\theta) = \frac{1}{\sqrt{2}}\overline{F_0(\theta)}$ and assume that $m_0(\theta)$ is 0 at $\theta = \pi$ and that this 0 has order $q \geq 1$. Then $|m_0(\theta)|^2 = 1 + O(|\theta|^{2q})$ and $m_1(\theta) = O(|\theta|^q)$ when θ tends to 0. Under these conditions, we say that F_0 is a low-pass filter and that F_1 is a high-pass filter, even though this terminology may not always be justified.

If these conditions are satisfied, the trend and the fluctuation around this trend of a signal X are defined, respectively, by $X_0 = T_0^* T_0(X)$ and $X_1 = T_1^* T_1(X)$.

The trend X_0 is twice as regular as X and, as such, can be subsampled by keeping only one point in two.

But this subsampling is furnished here by construction. Indeed, $T_0^* : l^2(2\mathbb{Z}) \to l^2(\mathbb{Z})$ is a partial isometry, and $T_0(X)$ constitutes the subsampling of the trend X_0. In the same way, $T_1(X)$ is the subsampling of the fluctuation X_1.

3.5. The time-scale algorithm of Mallat and the time-frequency algorithm of Galand.

It is amazing to reread Galand's thesis in the light of present understanding. Indeed, Galand's goal was to obtain finer and finer frequency resolutions by appropriately iterating the quadrature mirror filters. This is possible, however, only in the case of the ideal filters in our third example, and these ideal filters are unusable because they have an infinite impulse response. In spite of this criticism, we will return to Galand's point of view in Chapter 7, and it will lead us to wavelet packets.

Galand was thus looking for time-frequency algorithms, but his fundamental discovery, the quadrature mirror filters, was diverted from that end by Mallat, who intentionally used quadrature mirror filters to construct, using an hierarchical scheme, time-scale algorithms.

Mallat considers an increasing sequence $\Gamma_j = 2^{-j}\mathbb{Z}$ of nested grids that go from the "fine grid" $\Gamma_N (N \geq 1)$ to the "coarse grid" Γ_0. The signal to be analyzed has been sampled on the fine grid (we will come back to the sampling technique when studying the convergence problem), and our starting point is thus a sequence $f = f_0$ belonging to $l^2(\Gamma_n)$.

In addition, two quadrature mirror filters F_0 and F_1 are given. (We will see later what conditions they must satisfy.) These same filters will be used throughout the discussion.

We process the signal f by decomposing it into its trend and fluctuation. The trend is sampled on the next grid Γ_{n-1}; it represents a new signal that is decomposed again into trend and fluctuation. The fluctuations are never analyzed in this hierarchical scheme, and the algorithm follows a "herringbone" pattern.

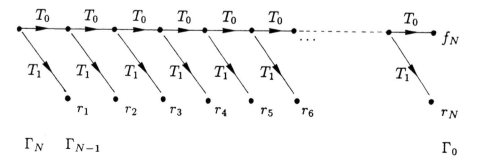

The input signal $f \in l^2(\Gamma_N)$ is finally represented by the sequence $r_1, \ldots, r_N$ of fluctuations and by the last trend $f_N \in l^2(\Gamma_0)$. The transformation that maps f onto the sequence $(r_1, \ldots, r_N, f_N)$ is clearly orthogonal, and the inverse is immediately calculated, based on the perfect reconstruction property of the quadrature mirror filters.

The significance of Mallat's algorithm stems from the following observation: For appropriate choice of the filters F_0 and F_1, there are numerous cases where the fluctuations $r_1, \ldots, r_N$ are, at different steps, extremely small. Coding the signal thus comes down to coding the last trend f_N as well as those coefficients of the fluctuations that are above the threshold fixed by the quantization. Notice that the last trend contains 2^{-N} times less data than the input signal and that the gain is appreciable.

3.6. Trends and fluctuations with orthonormal wavelet bases.

We propose to describe the asymptotic behavior of Mallat's algorithm when the number of stages N tends to infinity. In order to do this, it is first necessary to present the continuous version of this algorithm. This involves orthonormal wavelet bases in the following "complete form," which means that we have a wavelet plus a multiresolution analysis. This will be explained.

We begin with a function $\varphi(x)$ belonging to $L^2(\mathbb{R})$ that has the following property

$$(3.6) \qquad \varphi(x - k), \qquad k \in \mathbb{Z}, \quad \text{is an orthonormal sequence in } L^2(\mathbb{R}).$$

Let V_0 denote the closed linear subspace of $L^2(\mathbb{R})$ generated by this sequence. More generally, define V_j in terms of V_0 by simply changing scale, that is to say,

$$(3.7) \qquad\qquad f(x) \in V_0 \iff f(2^j x) \in V_j,$$

for functions $f \in L^2(\mathbb{R})$.

The other hypotheses are these: The V_j, $j \in \mathbb{Z}$, form a nested sequence; their intersection $\cap_{-\infty}^{\infty} V_j$ reduces to $\{0\}$; and their union $\cup_{-\infty}^{\infty} V_j$ is dense in $L^2(\mathbb{R})$.

We then write $\varphi_{j,k}(x) = 2^{j/2} \varphi(2^j x - k)$, $j, k \in \mathbb{Z}$, and define the trend f_j, at scale 2^{-j}, of a function $f \in L^2(\mathbb{R})$ by

$$f_j(x) = \sum_k \langle f, \varphi_{j,k} \rangle \varphi_{j,k}(x).$$

The fluctuations—or details in the case of an image—are denoted by $d_j(x)$ and defined by $d_j(x) = f_{j+1}(x) - f_j(x)$.

To analyze these details further, we let W_j denote the orthogonal complement of V_j in V_{j+1}, so that $V_{j+1} = V_j \oplus W_j$. Then there exists at least one function ψ belonging to W_0 such that $\psi(x - k)$, $k \in \mathbb{Z}$, is an orthonormal basis of W_0. This function $\psi(x)$, called the mother wavelet, has the following properties:

$$(3.8) \qquad\qquad 2^{j/2} \psi(2^j x - k) = \psi_{j,k}(x), \qquad j, k \in \mathbb{Z},$$

is an orthonormal basis for $L^2(\mathbb{R})$ and, more precisely, for all $j \in \mathbb{Z}$, we have

$$(3.9) \qquad\qquad d_j(x) = \sum_k \beta_{j,k} \psi_{j,k}(x).$$

The details, at a given scale, are thus linear combinations of the "elementary fluctuations," which are the wavelets related to that scale.

This analysis technique will be discussed again in the next chapter under the name "multiresolution analysis."

Given two functions $\varphi(x)$ and $\psi(x)$, called, respectively, "the father and mother wavelet," it is possible to define two quadrature mirror filters F_0 and F_1 by way of the operators $T_0 = DF_0$ and $T_1 = DF_1$. This is done by relating the approximation of the function space $L^2(\mathbb{R})$ that is given by the nested sequence of subspaces V_j to the approximation of the real line $\mathbb{R}$ that is given by the nested sequence of grids $\Gamma_j = 2^{-j}\mathbb{Z}$.

To do this, we consider that the function $\varphi_{j,k}(x) = 2^{j/2}\varphi(2^j x - k)$ is centered around the point $k2^{-j}$, which would be the case if $\varphi(x)$ was an even function. We associate the point $k2^{-j}$ with the function $\varphi_{j,k}$. This gives a correspondence between Γ_j and the orthonormal basis $\{\varphi_{j,k}, k \in \mathbb{Z}\}$ of V_j. At the same time, $l^2(\Gamma_j)$ is identified isometrically with V_j.

To define the operator $T_0 : l^2(\Gamma_{j+1}) \to l^2(\Gamma_j)$, it is sufficient to define its adjoint $T_0^* : l^2(\Gamma_j) \to l^2(\Gamma_{j+1})$. This adjoint T_0^* is a partial isometry. It is constructed by starting with the isometric injection $V_j \subset V_{j+1}$ and by identifying V_j with $l^2(\Gamma_j)$ and V_{j+1} with $l^2(\Gamma_{j+1})$, as explained above.

The orthonormal basis $\psi_{j,k}$ of W_j allows us to identify W_j with $l^2(\Gamma_j)$ in the same way. The isometric injection of $W_j \subset V_{j+1}$, interpreted with this identification, becomes the partial isometry

$$T_1^* : l^2(\Gamma_j) \to l^2(\Gamma_{j+1}).$$

Finally, the couple (φ, ψ) is represented by the couple (T_0, T_1) or, which amounts to the same thing, by the pair (F_0, F_1) of the two quadrature mirror filters. This crucial observation is due to Mallat.

Mallat also posed the converse problem: Given two quadrature mirror filters F_0 and F_1, is it possible to associate with them two functions φ and ψ having properties (3.6), (3.7), and (3.8)?

Although the converse is incorrect in general, it is correct in numerous cases, and this led to the construction of Daubechies's wavelets.

Our first and third examples of quadrature mirror filters show that the converse is generally false. There are no functions φ and ψ behind these numerical algorithms.

The second example is related to the Haar system. The function $\varphi(x)$ is the index function of $[0,1)$, and V_j is composed of step functions that are constant on each interval $[k2^{-j}, (k+1)2^{-j})$, $k \in \mathbb{Z}$.

The fourth example leads to "Shannon's wavelets." The function $\varphi(x)$ is the cardinal sine defined by $\frac{\sin \pi x}{\pi x}$.

Finally, the last example is more interesting, because both of the functions $\varphi(x)$ and $\psi(x)$ belong to the Schwartz class $\mathcal{S}(\mathbb{R})$, which consists of the infinitely differentiable functions that decrease rapidly at infinity.

In the next section, we are going to arrive at the functional analysis (that is, the continuous case) by passing to the limit in the discrete algorithms. To do this, we will give sufficient conditions on $m_0(\theta)$ to construct a multiresolution analysis starting with two quadrature mirror filters.

3.7. Convergence to wavelets.

In order to "restrict" a very irregular function $f(x)$ belonging to $L^2(\mathbb{R})$ to a grid $\Gamma = h\mathbb{Z}$, $h > 0$, it is necessary to filter the function before sampling. This filtering ought to be done according to specific rules so that, in the event $f(x)$ is more regular than anticipated, f can be reconstructed from the sampled version by interpolation.

Today we know exactly what needs to be done, and the technique of sampling is a direct consequence of Shannon's work.

We filter f by forming the convolution $f * g_h$, where $g_h(x) = h^{-1}g(h^{-1}x)$ and where $g(x)$ and its Fourier transform $\hat{g}(\xi)$ are chosen so that

(3.10) $\quad$ $g(x)$ is in the class C^r and $g(x), g'(x), \ldots, g^{(r)}(x)$ all decrease rapidly at infinity,

(3.11) $$\int_{-\infty}^{\infty} g(x)dx = 1, \quad \text{and}$$

(3.12) $$\hat{g}(2k\pi) = 0 \quad \text{if} \quad k \in \mathbb{Z}, \quad k \neq 0.$$

One can then restrict the filtered signal to the grid $h\mathbb{Z}$.

We assume that these conditions are satisfied throughout the discussion, and we begin by reconsidering Mallat's "herringbone" algorithm. Start by fixing f in $L^2(\mathbb{R})$, and sample f on the grid Γ_N using the preconditioning filter $f * g_N$, where $g_N(x) = 2^N g(2^N x)$.

We wish to study the asymptotic behavior of Mallat's algorithm as N tends to infinity. The limit we are looking for is defined as follows: Fix the index j of the grid Γ_j. (Starting with Γ_0, we will look at Γ_1, Γ_2, etc. ...) Then we seek the (simple) limits of the sequences

$$f_N(k), r_N(k), r_{N-1}(2^{-1}k), \ldots, r_{N-j}(2^{-j}k), \ldots$$

when N tends to $+\infty$, j and k being fixed. Refer to the "herringbone" scheme for the definitions of $f_N, r_N, r_{N-1}, \ldots$ Here are the results [6].

THEOREM 3.2. *Assume that the impulse responses of the quadrature mirror filters F_0 and F_1 decrease rapidly at infinity and that the transfer function $F_0(\theta)$ of F_0 satisfies $F_0(0) = \sqrt{2}$, $F_0(\theta) \neq 0$ if $-\frac{\pi}{2} \leq \theta \leq \frac{\pi}{2}$. Then Mallat's "herringbone" algorithm, applied to $f * g_N$, as indicated above, converges to the analysis of f in an orthonormal wavelet basis.*

More precisely,

$$\lim_{N \to +\infty} f_N(k) = \int_{-\infty}^{\infty} f(x) \,\overline{\varphi(x-k)}\, dx,$$

$$\lim_{N \to +\infty} r_N(k) = \int_{-\infty}^{\infty} f(x) \,\overline{\psi(x-k)}\, dx, \dots,$$

$$\lim_{N \to +\infty} r_{N-j}(2^{-j}k) = \int_{-\infty}^{\infty} f(x) \,\overline{\psi_{j,k}(x-k)}\, dx.$$

The functions $\varphi(x)$ and $\psi(x)$ are, respectively, "the father and mother" of the orthonormal wavelet basis, as explained in the preceding section.

A very beautiful application of this result (to which we return in Chapter 4) is the construction of the celebrated bases of Daubechies.

3.8. The wavelets of Daubechies.

These wavelets depend on an integer $N \geq 1$ that defines the support of the functions $\varphi(x)$ and $\psi(x)$, namely, $[0, 2N - 1]$, as well as the Hölder regularity of these functions: $\varphi(x)$ and $\psi(x)$ belong to C^r, where $r = r(N)$ and $\lim_{N \to +\infty} N^{-1} r(N) = \gamma > 0$. The value of γ is about $1/5$; this implies that, if a wavelet $\psi(x)$ is to have 10 continuous derivatives, the length of its support must be about 50.

The functions $\varphi(x)$ and $\psi(x)$, which ought to be written as φ_N and ψ_N, are, respectively, the father and mother of the orthonormal wavelet basis.

To construct this orthornormal basis, Daubechies applies the method of the last section. One starts with the trigonometric sum

$$P_N(t) = 1 - c_N \int_0^t (\sin u)^{2N-1} du = \sum_{|k| \leq 2N-1} \gamma_k e^{ikt}$$

with the constant $c_N > 0$ chosen so that $P_N(\pi) = 0$. There exists (at least one) finite trigonometric sum $m_0(t) = \frac{1}{\sqrt{2}} \sum_0^{2N-1} h_k e^{-ikt}$ such that $|m_0(t)|^2 = P_N(t)$ and $m_0(0) = 1$. The coefficients h_k constitute the impulse response of the filter F_0 and are real.

Under these conditions we know from Theorem 3.2 that the functions φ and ψ exist and that they form a multiresolutional analysis. We now use these results to construct φ and ψ explicitly. The function φ, which we seek to construct, must be a solution of the functional equation

(3.13) $$\varphi(x) = \sqrt{2} \sum_0^{2N-1} h_k \varphi(2x - k) \quad \text{and} \quad \int_{-\infty}^{\infty} \varphi(x) dx = 1.$$

This functional equation follows from the inclusion $V_0 \subset V_1$ and the fact that $\varphi(x - k)$, $k \in \mathbb{Z}$, is an orthonormal basis for V_0 and $\sqrt{2}\varphi(2x - k)$, $k \in \mathbb{Z}$, is an orthonormal basis for V_1.

This functional equation leads to

$$(3.14) \qquad \hat{\varphi}(\xi) = m_0(\xi/2)m_0(\xi/4)\ldots m_0(\xi/2^j)\ldots,$$

and the principal difficulty in this construction is to show that one has $\hat{\varphi}(\xi) = O(|\xi|^{-m})$ at infinity, where $m = m(N)$ tends to infinity with N.

On the other hand, it is almost obvious that the support of $\varphi(x)$ is in $[0, 2N - 1]$.

The fact that $\varphi(x - k)$, $k \in \mathbb{Z}$, is an orthonormal sequence is a direct consequence of Theorem 3.2 and the fact that $m_0(t) \neq 0$ on $[-\pi/2, \pi/2]$.

To determine the Fourier transform $\hat{\psi}(\xi)$ of the wavelet $\psi(x)$, we write $m_1(t) = e^{i(1-2N)t}\,\overline{m_0(t + \pi)}$. Then

$$(3.15) \quad \hat{\psi}(\xi) = m_1(\xi/2)\hat{\varphi}(\xi/2) = m_1(\xi/2)m_0(\xi/4)m_0(\xi/8)\ldots m_0(\xi/2^j)\ldots$$

and the support of $\psi(x)$ is the interval $[0, 2N - 1]$.

If $N = 1$, $\varphi(x)$ is the index function of $[0,1)$, while $\psi(x) = 1$ on $[0,1/2)$, -1 on $[1/2,1)$ and 0 elsewhere. The orthonormal basis $2^{j/2}\psi(2^j x - k)$, $j, k \in \mathbb{Z}$, is then the Haar system.

3.9. Conclusions.

The functions ψ_N used by Daubechies to construct the orthonormal bases named for her are new "special functions." These "special functions" never appeared in previous work, and their only definition is provided by (3.14) and (3.15). This means that the detour by way of quadrature mirror filters and the corresponding transfer functions was indispensable. In other words, it never would have been possible to discover Duabechies's wavelets by trying to solve directly the existence problem: Is there, for each integer $r \geq 0$, a function $\psi(x)$ of class C^r such that $2^{j/2}\psi(2^j x - k)$, $j, k \in \mathbb{Z}$, is an orthonormal basis of $L^2(\mathbb{R})$?

The quadrature mirror filters will be used again in the algorithms for numerical image processing, which we describe in the next chapter.

Bibliography

[1] E. H. ADELSON, R. HINGORANI, AND E. SIMONCELLI, *Orthogonal pyramid transforms for image coding*, SPIE, Visual Communications and Image Processing II, Cambridge, MA, October 27–29, 1987, Vol. 845, pp. 50–58.

[2] I. DAUBECHIES, *Orthonormal bases of compactly supported wavelets*, Comm. Pure Appl. Math., 41 (1988), pp. 909–996.

[3] D. ESTEBAN AND C. GALAND, *Application of quadrature mirror filters to split band voice coding systems*, International Conference on Acoustics, Speech, and Signal Processing, Washington, DC, May 1977, pp. 191–195.

[4] C. GALAND, *Codage en sous-bandes: théorie et applications à la compression numérique du signal de parole*, Thesis, University of Nice, Nice, France, March 1983.

[5] S. MALLAT, *A theory for multiresolution signal decomposition: The wavelet representation*, IEEE Trans. Pattern Anal. Machine Intell., 11 (1989), pp. 674–693.

[6] Y. MEYER AND F. PAIVA, *Convergence de l'algorithme de Mallat*, preprint, CEREMADE, University of Paris-Dauphine, Paris, France.

CHAPTER **4**

Pyramid Algorithms for Numerical Image Processing

4.1. Introduction.

Experts in image processing agree on the following point: An image contains important information in a wide range of scales, and this information is often independent from one scale to another.

Marr wrote in *Vision* [9, p. 51]: "Although the basic elements in our image are the intensity changes, the physical world imposes on these raw intensity changes a wide variety of spatial organizations, roughly independently at different scales." And we read on page 54: "Intensity changes occur at different scales in an image, and so their optimal detection requires the use of operators of different sizes."

In [2], Adelson, Hingorani, and Simoncelli used the same language: "Images contain information at all scales."

Cartography illustrates this concept very well. Maps contain different information at different scales. For example, it is impossible to plan a trip to visit the Roman churches at Charente and Poitou using the map of France found on a globe of the earth. Indeed, the villages where these churches are found do not appear on the global representation.

Cartographers have developed conventions for dealing with geographic information by partitioning it into independent categories that correspond to the different scales of a department, a region, a country, a continent, and the whole globe. These categories are not entirely independent, and the more important features existing at a given scale are repeated at the next larger scale. Thus, it is sufficient to specify the relations between information given at two adjacent scales in order to define unambiguously the embedding of the different representations at different scales. Naturally these embedding relations (such as which department belongs to which province, and which province belongs, in turn, to which country, and so on. . .) are available to us from our knowledge of geography; however, they could be discovered by merely examining the maps.

We can see from this example the fundamental idea of representing an image by a tree. In the cartographic case, the trunk would be the map of the world. By traveling toward the branches, the twigs, and the leaves, we reach successive maps that cover smaller regions and give more details, details that do not appear at lower levels.

45

To interpret this cartographic representation using the pyramid algorithm, it will be necessary to reverse the roles of top and bottom, since the pyramid algorithm progresses from "fine to coarse." In cartography, usage and certain conventions determine which details are deleted in going from one scale to another and which "coherent structures" (see Chapter 10) persist across a succession of scales.

In this chapter, we are going to describe the pyramid algorithms of Burt and Adelson, as well as two important modifications derived from them. The purpose of these algorithms is to provide an automatic process, in the context of digital imagery, to calculate the image at scale 2^{j+1} from the image at scale 2^j. If the original image corresponds to a fine grid with 1024×1024 points, the pyramid algorithm first yields a 512×512 image, then one 256×256, next a 128×128 image, and so on until reaching the absurd (in practice) 1×1 limit. The interest in pyramid algorithms derives from their iterative structure, which uses results from a given scale, 2^j, to move to the next scale, 2^{j+1}.

Returning to our cartographic example, suppose that we already have maps of the French Departments at a scale of 1 to 200,000. Then it is of no value to refer to the new satellite images in order to construct a map of France at a scale of 1 to 2,000,000. The information needed to make this new map is already contained in the maps of the Departments. The point is that one uses judiciously the work already done without going back to the raw data. We have just outlined the general philosophy of the pyramid algorithms without, however, describing the algorithms that are used to change scale. How, starting with a very precise representation of the Brittany coast at a scale of 1 to 200,000, can we arrive at a more schematic description at a scale of 1 to 2,000,000 without smoothing or softening too much the myriad details and roughness that characterize the Brittany coastline?

The pyramid algorithms of Burt and Adelson and their variants (orthogonal and bi-orthogonal pyramids) deal with this type of problem. In all cases, this will involve calculating (at each scale) an approximation f_j of a given image by using an interative algorithm to go from one scale to the next.

4.2.   The pyramid algorithms of Burt and Adelson.

For the rest of the discussion, $\Gamma_j = 2^{-j}\mathbb{Z}^2$ will denote the sequence of nested grids used for image processing. It often happens that the image is bounded by the unit square, in which case we will speak of a 512×512 image to indicate that $j = 9$; similarly, a 1024×1024 image will correspond to $j = 10$.

At this point, we are working with images that are already digitized and appear as numerical functions. The raw image that provided these digital images will be, for us, a function $f(x, y)$. This function can be very irregular, either because of noise or because of discontinuities in the image. For example, discontinuities can be due to the edges of objects in the image.

The sampled images f_j are defined on the corresponding grids $\Gamma_j = 2^{-j}\mathbb{Z}^2$. These sampled images are obtained from the original physical image, that is, $f(x, y)$, by the restriction operators $R_j : L^2(\mathbb{R}^2) \rightarrow l^2(\Gamma_j)$. These operators R_j

will be defined in the following pages. They are the same type as those used in numerical analysis to discretize an irregular function or distribution.

The fundamental discovery of Burt and Adelson is the existence of restriction operators R_j with the property that, for all initial images f, the sampled images $f_j = R_j(f)$ are related to each other by extremely simple causality relations.

These causality relations, of the type "fine to coarse," allow f_{j-1} to be calculated directly from f_j without having to go back to the original physical image $f(x, y)$.

In order to define the restriction operators R_j, we first consider the case of a grid given by $x = hk$, $y = hl$, where $h > 0$ is the sampling step and $(k, l) \in \mathbb{Z} \times \mathbb{Z}$.

Very irregular functions should not be sampled directly and, therefore, the image must be smoothed before it is discretized. This leads to the following classic scheme

where F is a low-pass filter prior to the sampling E.

To determine the characteristics of the filter F, first consider the special case, $f(x, y) = \cos(mx + ny + \varphi)$, $m, n \in \mathbb{N}$. To sample this function correctly on $h\mathbb{Z}^2$, the Nyquist condition must be satisfied. This means that h must be less than $\min\{\pi/m, \pi/n\}$ if we wish to be able to reconstruct $f(x, y)$. Looked at from the other side, the Nyquist condition says that sampling on $h\mathbb{Z}^2$ will lose all information at frequencies higher than π/h. For the case at hand, the Nyquist condition comes down to suppressing, through the action of the filter F, all the frequencies in $f(x, y)$ that are greater than π/h. This is done by smoothing the signal through convolution with $\frac{1}{h^2} g\left(\frac{x}{h}, \frac{y}{h}\right)$, where $g(x, y)$ is a sufficiently regular function concentrated around zero.

The filtering/sampling scheme maps the physical image $f(x, y)$ onto a numerical image defined by

$$(4.1) \qquad c(k, l) = \frac{1}{h^2} \iint g\left(k - \frac{x}{h}, l - \frac{y}{h}\right) f(x, y) dx \, dy.$$

By writing $\varphi(x, y) = \overline{g(-x, -y)}$ and $\varphi_h(x, y) = \frac{1}{h^2} \varphi\left(\frac{x}{h}, \frac{y}{h}\right)$, we have

$$(4.2) \qquad c(x, y) = \langle f, \varphi_h(\cdot - kh, \cdot - lh)\rangle$$

where $\langle u, v \rangle = \iint u(x, y)\overline{v(x, y)} dx \, dy$ and where $\cdot$ denotes the (dummy) variable of integration.

The *extension operator* enables us to extend a sequence $c(k, l)$ defined on $h\mathbb{Z}^2$ to a regular function on $\mathbb{R}^2$. In this sense, it is inverse to the filtering/sampling operation. We define the extension operator to be the *adjoint of the restriction operator* so it is given by

$$(4.3) \qquad c(k, l) \rightarrow \sum \sum c(k, l)\varphi(h^{-1}x - k, h^{-1}y - l).$$

This is an *interpolation operator*.

The simplest examples are given by the spline functions. We consider the one-dimensional case to simplify the notation. If we let T be the "triangle" function $T(x) = \sup(1 - |x|, 0)$, then (4.3) yields the usual piecewise linear interpolation of a discrete sequence. A second choice is given by $\varphi = T * T$, which is the basic cubic spline.

Returning to the general case, we ask that the operator $P_h R_h$, composed of the restriction operator followed by the extension operator, have the property, that for all functions $f \in L^2(\mathbb{R}^2)$,

$$(4.4) \qquad P_h R_h(f) \to f, \quad \text{in the quadratic mean, when } h \text{ tends to } 0.$$

By assuming, for example, that $\varphi(x)$ is a continuous function that decreases rapidly at infinity, it is easy to show that (4.4) is equivalent to the condition $P_h R_h(1) = 1$, where 1 represents the function identically equal to 1. And this is equivalent to

$$(4.5) \qquad |\hat{\varphi}(0,0)| = 1, \qquad \hat{\varphi}(2k\pi, 2l\pi) = 0 \quad \text{if} \quad (0,0) \neq (k,l) \in \mathbb{Z}^2.$$

In what follows, we assume that $\iint \varphi(x,y)dx\, dy = 1$, after possibly multiplying φ by a constant of modulus 1.

We return to the fundamental problem posed by Burt and Adelson. Thus we consider the nested sequence $\Gamma_j = 2^{-j}\mathbb{Z}^2$. These grids become finer when j tends toward $+\infty$ and coarser when j tends toward $-\infty$.

We begin with a function $\varphi(x,y)$ that is continuous on $\mathbb{R}^2$ and decreases rapidly at infinity. We also assume, as above, that $\hat{\varphi}(0,0) = 1$. Denote by R_j and P_j the restriction and extension operators associated with this choice of φ and the grid Γ_j.

Burt and Adelson's basic idea is that, for certain choices of the function φ, the different "sampled images" $R_j(f) = f_j$ derived from the same "physical image" f are necessarily related by extremely simple "causality relations." The dynamic of these relations is from "fine to coarse," which means that a function defined on a fine grid is mapped to one on a coarse grid. To make these causality relations explicit, we denote by T_j the operators that will eventually be defined by these relations, that is, by $T_j(f_j) = f_{j-1}$, where $f_j = R_j(f)$ and $f_{j-1} = R_{j-1}(f)$. We can summarize all this with the two conditions:

$$(4.6) \qquad\qquad\qquad T_j : l^2(\Gamma_j) \to l^2(\Gamma_{j-1}),$$

$$(4.7) \qquad\qquad\qquad R_{j-1} = T_j R_j.$$

One might naïvely think that the operator T_j can be defined by inverting the operator R_j in (4.7). But the operator R_j is a smoothing operator, and its inverse is not defined. In terms of images, it is not possible to go from a blurred image back to the original image.

Thus we cannot solve (4.7) by elementary algebra. On the other hand, once R_j is restricted to an appropriate closed subspace V_j of $L^2(\mathbb{R}^2)$, $R_j : V_j \to l^2(\Gamma_j)$ becomes, in certain cases, an isomorphism. Then we can solve (4.7) directly.

Burt and Adelson asked how to determine the functions φ such that (4.6) and (4.7) are satisfied. Stated this way, the problem is very difficult, for most of the usual choices of smoothing functions do not have these properties. To resolve this difficulty, Burt and Adelson proceeded the other way around; that is to say, they sought to construct φ from the operators T_j.

For this it is necessary to derive some consequences of (4.7). The first is that the operator $T_0 : l^2(\mathbb{Z}^2) \to l^2(2\mathbb{Z}^2)$ can be written as $T_0 = DF_0$, where $F_0 : l^2(\mathbb{Z}^2) \to l^2(\mathbb{Z}^2)$ is a filter operator and where $D : l^2(\mathbb{Z}^2) \to l^2(2\mathbb{Z}^2)$ restricts a function defined on $\mathbb{Z}^2$ to $2\mathbb{Z}^2$. D is the decimation operator, which we have already encountered in Chapter 3. The fact that T_0 has this form is a consequence of the fact that T_0 commutes with all even translations.

Thus, if $X = (x(k))_{k \in \mathbb{Z}^2}$, we can write

$$(4.8) \qquad T_0(X)(2k) = \sum_{l \in \mathbb{Z}^2} \omega(2k - l)x(l),$$

where $\omega(k)$ is the impulse response of the filter F_0. For convenience, we assume (as Burt and Adelson did) that $\omega(k)$ is real.

If we apply T_0 to $x(k) = \int f(x)\overline{\varphi(x - k)}dx$, then from (4.7) we get $\frac{1}{4} \int f(x)\varphi\left(\frac{x}{2} - k\right)dx$. Condition (4.8) is thus equivalent to

$$(4.9) \qquad \varphi(x, y) = 4 \sum_k \sum_l \omega(k, l)\varphi(2x + k, 2y + l).$$

Note that the meaning of k changed in (4.9): We had $k \in \mathbb{Z}^2$ in the preceding formulas, while k and l now belong to $\mathbb{Z}$.

For obvious reasons, Burt and Adelson were particularly interested in filters with finite length. This means that $\omega(k, l)$ is zero if $|k| > N$ and $|l| > N$ for some N.

By taking Fourier transforms of both sides, (4.9) becomes

$$(4.10) \qquad \hat{\varphi}(\xi, \eta) = m_0\left(\frac{\xi}{2}, \frac{\eta}{2}\right)\hat{\varphi}\left(\frac{\xi}{2}, \frac{\eta}{2}\right),$$

where

$$(4.11) \qquad m_0(\xi, \eta) = \sum_k \sum_l \omega(k, l)e^{i(k\xi + l\eta)}.$$

Recall that $\hat{\varphi}(0, 0) = 1$. By iterating (4.10) and passing to the limit, it becomes

$$(4.12) \qquad \hat{\varphi}(\xi, \eta) = m_0\left(\frac{\xi}{2}, \frac{\eta}{2}\right)m_0\left(\frac{\xi}{4}, \frac{\eta}{4}\right) \cdots m_0\left(\frac{\xi}{2^j}, \frac{\eta}{2^j}\right) \cdots$$

The second consequence that we derive from (4.7) is that all of these conditions (for different j) are in fact equivalent. This can be seen by making the change of variables $x \mapsto 2^j x$ and $y \mapsto 2^j y$ in (4.9), and integrating both sides against $f(x, y)$. We then have

$$(4.13) \qquad R_{j-1}(f)(2^{-j+1}k) = \sum_{l \in \mathbb{Z}^2} \omega(2k - l)R_j(f)(2^{-j}l).$$

In other words, under our assumptions, the operators T_j are defined by

$$(4.14) \qquad T_j(X)(2^{-j+1}k) = \sum_{l \in \mathbb{Z}^2} \omega(2k - l)x(2^{-j}l)$$

when $X = (x(2^{-j}l))$ belongs to $l^2(\Gamma_j)$. The point is that the sequence $\omega(k)$, $k \in \mathbb{Z}^2$, is the same for all the operators T_j. Working backwards, Burt and Adelson began with a finite sequence of coefficients $\omega(k, l)$ such that $\sum_k \sum_l \omega(k, l) = 1$. They defined $m_0(\xi, \eta)$ by (4.11) and then $\hat{\varphi}$ by (4.12). Then the first question to arise is whether the second member of (4.12) defines a square-integrable function. If this is the case, we call this function $\hat{\varphi}$, we define the restriction operators R_j in terms of the Fourier transform of this function, and we define the transition operators T_j by (4.14). Then $R_{j-1} = T_j R_j$ for all $j \in \mathbb{Z}$.

4.3.   Examples of pyramid algorithms.

Before continuing the presentation of the Burt and Adelson algorithms, we give examples of functions φ that illustrate both the existence and the nonexistence of the transition operators. Conversely, we give examples of sequences $\omega(k, l)$ illustrating the existence and nonexistence of the associated function φ.

We begin with two examples where the transition operators do not exist. Suppose that $\varphi(x, y) = \frac{1}{\pi}e^{-x^2-y^2}$. Then there are no transition operators because (4.10) implies that $m_0(\xi, \eta) = \exp\left(-\frac{3}{4}(\xi^2 + \eta^2)\right)$, which is clearly not 2π-periodic in ξ and η. In the same way, the transition operators do not exist if $\varphi(x, y) = \frac{1}{4}\exp(-|x| - |y|)$. One senses, justifiably, that the existence of transition operators is exceptional.

Here, however, is an example where the operators do exist. To simplify the discussion, this example (the spline functions) is presented in dimension one. Let $m \geq 0$ be an integer, and let $\varphi(x)$ be the convolution product $\chi * \cdots * \chi$. Here there are m products and $m + 1$ terms, and χ is the characteristic function of the interval $[0, 1]$. Then $\hat{\varphi}(\xi) = \left((e^{-i\xi} - 1)/ - i\xi\right)^{m+1}$, and (4.10) is satisfied with $m_0(\xi) = \left((1 + e^{-i\xi})/2\right)^{m+1}$, which is indeed 2π-periodic.

We now proceed the other way; we start with a sequence of transition operators (T_j), which is a sequence $\omega(k, l)$, $k, l \in \mathbb{Z}$, and we propose to reconstruct φ. All the examples that we consider are constructed with separable sequences $\omega(k, l)$, that is, sequences of the form $\omega(k)\omega(l)$. The associated function $\varphi(x, y)$ will then necessarily be of the form $\varphi(x)\varphi(y)$.

For the first example take $\omega(k) = 0$ if $k \neq 0$ and $\omega(0) = 1$. In this case the function φ defined by (4.12) is the Dirac measure at 0 and the restriction operators $R_j : L^2(\mathbb{R}^2) \to l^2(\Gamma_j)$ are no longer defined.

In the second example take $\omega(k) = 0$ except for $k = \pm 1$, and $\omega(\pm 1) = 1/2$. From this we can deduce that $m_0(\xi) = \cos(\xi)$ and $\varphi(x) = 1/2$ on the interval $[-1, 1]$ and 0 elsewhere. This choice for φ, which is (for the moment) perfectly reasonable, will be excluded when we introduce the concept of multiresolution analysis.

Burt and Adelson proposed a very original function for ω, and this will be our third example. Take $\omega(0) = 0.6$; $\omega(\pm 1) = 0.25$; $\omega(\pm 2) = -0.05$; and

$w(k) = 0$ for $|k| \geq 3$. The corresponding function $\varphi(x)$ is continuous, its support is $[-2, 2]$, and it resembles $C \exp(-c|x|)$, $C > 0$, $c > 0$, on this interval. This is why the corresponding algorithm is called a Laplacian pyramid. We shall see this example again when we introduce bi-orthogonal wavelets at the end of the chapter.

The purpose of our last example is to show that the existence of the function φ, defined by (4.12), is not a stable property, even in the simplest cases. In fact, we limit our discussion to functions $w(k)$ that are zero except at $k = 0$ and $k = -1$, and here $w(0) = p$, $w(-1) = q$, $0 < p < 1$, $0 < q < 1$, $p + q = 1$. Then the choice $p = q = 1/2$ leads to a function φ that is the characteristic function of $[0, 1]$. All other choices imply that the mathematical object on the right side of (4.12) is the Fourier transform of a probability measure μ that is singular with respect to the Lebesque measure. The support of this probability measure μ is the interval $[0, 1]$. This measure is defined by the following property: If I is a dyadic interval in $[0, 1]$ and if I' is that left half of I and I'' is the right half of I, then $\mu(I') = p\mu(I)$ and $\mu(I'') = q\mu(I)$.

We drop for the moment the problem of choosing an optimal filter $w(k)$, $k \in \mathbb{Z}^2$. Indeed, such a choice must take into consideration the overall objective. Burt and Adelson's objective was image compression. We are going to present their compression algorithm in the next section. After that we will return to the problem of choosing $w(k)$.

4.4. Pyramid algorithms and image compression.

Image compression is one of the uses of the pyramid algorithms. Burt and Adelson's algorithm, which we describe in this section, will later be compared with other algorithms (orthogonal pyramids and bi-orthogonal wavelets) that perform better. All of the pyramid algorithms act on images that are already sampled and never on the original physical image. In other words, the function φ we have tried to construct using the sequence $w(k)$ is never used. Then why have we worked to learn its properties? The answer is given at the end of §4.6, but we can indicate the idea here. We want, on reaching the summit of the pyramid, to have a moderate, softened image and not one that is chaotic and noisy; this is precisely where the regularity of the function φ comes into play.

The Burt and Adelson pyramid algorithms use only the transition operators $T_j : l^2(\Gamma_j) \rightarrow l^2(\Gamma_{j-1})$. All of these operators are the same, except for a change of scale; therefore, we are going to assume that $j = 0$.

The discussion of the algorithm begins with the definition of the trend and the fluctuations around this trend for a sequence f belonging to $l^2(\Gamma_0)$. This trend cannot be $T_0(f)$ because it "lives in a different universe" and cannot be compared to f. To define the trend, it is necessary to leave the coarse grid $2\Gamma_0$, where $T_0(f)$ is defined, and return to the fine grid Γ_0, where f is defined. This is done by using the adjoint operator $T_0^* : l^2(2\Gamma_0) \rightarrow l^2(\Gamma_0)$, and the trend of f is defined by $T_0^* T_0(f)$.

We clearly want the trend of a very regular function to coincide with that function. This leads to the requirements that $T_0^* T_0(1) = 1$ and, more generally, that $T_0^* T_0(P) = P$ for all polynomials P of degree less than or equal to N, for some fixed integer N. This condition is equivalent to the following: The function $m_0(\xi, \eta)$, defined by (4.11), must vanish, along with all of its derivatives of order less than or equal to N, at points $(\varepsilon_1 \pi, \varepsilon_2 \pi)$, $\varepsilon_1, \varepsilon_2 \in \{0, 1\}$, with the exception of the origin. At the origin, one must have

$$|m_0(\xi, \eta)|^2 = 1 + O(|\xi| + |\eta|)^{N+1}.$$

The price to pay is naturally the length of the filter $w(k, l)$ that must be used to satisfy these conditions. This length is proportional to N. Another interesting observation is that the conditions we have just imposed on $m_0(\xi, \eta)$ imply, by (4.12), that $\hat{\varphi}(2k\pi, 2l\pi) = 0$ if $(k, l) \in \mathbb{Z}^2$ and $(k, l) \neq (0, 0)$. But this last condition is the same as (4.5), which, as we know, is necessary and sufficient to have $P_h R_h(f) \to f$ in the norm of $L^2(\mathbb{R}^2)$ when h tends to zero. Since T_0 is the discrete analogue of the restriction operator R_h, whereas T_0^* corresponds to the extension operator P_h, $T_0^* T_0$ is the "discrete approximation" operator analogous to the continuous approximation operator $P_h R_h$.

The fluctuation around the trend is $f - T_0^* T_0(f)$ when f belongs to $l^2(\Gamma_0)$. This fluctuation is zero whenever f is a polynomial of degree no greater than N, and one can easily deduce from this that the fluctuation will be very weak in all areas where the image is very regular. As we will see, this last property is the key to the success of the Burt and Adelson algorithm.

The trend, and the fluctuation around the trend, of a sequence f belonging to $l^2(\Gamma_j)$ are defined by a simple change of scale. The trend is $T_j^* T_j(f)$ and the fluctuation is $f - T_j^* T_j(f)$.

If the sequence in question, f, is the restriction to the grid Γ_j of a function F that is very regular in a some (open) region Ω, then

(4.15) $$|f - T_j^* T_j(f)| \leq C 2^{-(N+1)j}$$

at all the points of this region. Thus the Burt and Adelson algorithm becomes more effective as N increases.

To define the coding and compression algorithm of Burt and Adelson, we begin with the "fine grid" $\Gamma_m = 2^{-m} \mathbb{Z}^2$ and a numerical image f_m sampled on this fine grid. This numerical image is, in fact, the restriction to $\Gamma_m = 2^{-m} \mathbb{Z}^2$ of a physical image $f \in L^2(\mathbb{R}^2)$. This means that f_m is the restriction in the usual sense of the convolution product $f * g_m$, where $g_m(x, y) = 4^m g(2^m x, 2^m y)$. The properties of the function g were indicated in §4.2; in addition, we shall assume that the integral of g is equal to 1. However it is not necessary to "return to the physical image" f to use the algorithm.

Burt and Adelson replace f_m by the couple (trend, fluctuation). But the trend, which is given by $T_m^* T_m(f_m)$, is completely defined by $T_m(f_m)$. This amounts to saying that the trend is sufficiently regular that it can be coded by retaining one pixel in four. This coding is given by $T_m(f_m)$. In summary, Burt and Adelson code f_m with the couple $[T_m(f_m), f_m - T_m^* T_m(f_m)]$. Then the fluctuation, denoted by r_m, is not processed further. They write $f_{m-1} = T_m(f_m)$ and iterate the procedure. f_{m-1} is coded by (f_{m-2}, r_{m-1}), where $f_{m-2} = T_{m-1}(f_{m-1})$ and $r_{m-1} = f_{m-1} - T_{m-1}^* T_{m-1}(f_{m-1})$.

These substitutions are interesting for two reasons:

(4.16)   Going from f_j to f_{j-1} reduces the data one must
              deal with by a factor of four. Indeed, f_j is defined on
              Γ_j and f_{j-1} on Γ_{j-1}, which has one-fourth as many points.

(4.17)   In many cases, the fluctuations are small enough
              that they can be replaced by zero.

Condition (4.17) is satisfied in all the regions where the image exhibits a certain
regularity.

    If we suppose that the starting image f_m is bounded on a square of side 1,
then the algorithm is stopped on reaching the summit of the pyramid, which is
the grid Γ_0.

    The image f_m is coded by the sequence $(f_0, r_1, \ldots, r_m)$, where f_0, defined on
Γ_0, is a scalar and where the $r_j = (1 - T_j^* T_j) f_j$, $1 \le j \le m$, are the different
fluctuations.

    The diagram below gives a schematic description of the algorithm. The
columns correspond to the different grids Γ_j.

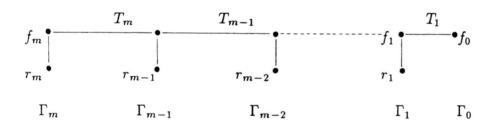

    Running the algorithm the other way, which is reconstructing f_m from the
code, is very easy. Begin with f_0 and r_1, and compute $f_1 = T_1^* f_0 + r_1$. In the
same way, reconstruct f_2 by $f_2 = T_2^* f_1 + r_2$, and continue until f_m is recovered.

    As we will see a little later, this algorithm is advantageous only if most
of the fluctuations are zero. Otherwise it is clearly disadvantageous because
information is wasted.

4.5.   Pyramid algorithms and multiresolution analysis.

Before leaving the first version of Burt and Adelson's algorithms, we are going to
describe a continuous version of it. The interplay between the discrete algorithms
and their continuous versions, which is implicit in the work of Burt and Adelson,
was made explicit by Mallat and the author.

    We consider the general case because dimension two plays no particular role
in the following definition.

    A *multiresolution analysis of* $L^2(\mathbb{R}^n)$ *is an increasing sequence of closed sub-
spaces* $(V_j)_{j \in \mathbb{Z}}$ *of* $L^2(\mathbb{R}^n)$ *having the following three properties:*

(4.18) $$\bigcap_{-\infty}^{\infty} V_j = \{0\}, \qquad \bigcup_{-\infty}^{\infty} V_j \quad \text{is dense in } L^2(\mathbb{R}^n),$$

(4.19)   *for all functions $f \in L^2(\mathbb{R}^n)$ and all integers $j \in \mathbb{Z}$*
              $f(x) \in V_0$ *is equivalent to* $f(2^j x) \in V_j$

(4.20)   *there exists a function $\varphi(x)$, belonging to V_0,*
              such that the sequence $\varphi(x - k)$, $k \in \mathbb{Z}^n$,
              is a Riesz basis for V_0.

Recall that if H is a Hilbert space, a *Riesz basis* $(e_j)_{j \in J}$ of H is, by definition, an isomorphic image $T : H \to H$ of a Hilbert basis $(f_j)_{j \in J}$ of H. (Note that T is not necessarily an isometry.) Then each vector $x \in H$ is decomposed uniquely in a series

(4.21) $$x = \sum_{j \in J} \alpha_j e_j \quad \text{where} \quad \sum |\alpha_j|^2 < \infty.$$

Furthermore, $\alpha_j = \langle x, e_j^* \rangle$, where $e_j^* = (T^*)^{-1}(f_j)$ is the dual basis of e_j; this dual basis is itself a Riesz basis.

We are going to measure the regularity of a multiresolution analysis; this will, in fact, be the regularity of the functions belonging to V_0. To measure this regularity, we introduce an integer r, which can take the values $0, 1, 2, \ldots$, and even $+\infty$.

We say that the multiresolution analysis $(V_j)_{j \in \mathbb{Z}}$ is r-regular if it is possible to choose the function $\varphi(x)$ in (4.20) so that, for all integers $m \geq 0$ and all $x \in \mathbb{R}^n$,

(4.22) $$|\partial^\alpha \varphi(x)| \leq C_m (1 + |x|)^{-m},$$

where $\alpha = (\alpha_1, \ldots, \alpha_n)$ is a multi-index satisfying $\alpha_1 + \cdots + \alpha_n \leq r$ and where $\partial^\alpha = (\partial/\partial x_1)^{\alpha_1} \ldots (\partial/\partial x_n)^{\alpha_n}$.

We return to the two-dimensional case. Here, a multiresolution analysis is, in a certain sense, a particular case of the pyramid algorithm. To see this, suppose that the function φ, which is defined by (4.12), has the following additional property: There exist two constants $C_2 \geq C_1 > 0$ such that for all scalar sequences $(\alpha_k)_{k \in \mathbb{Z}^2}$,

(4.23) $$C_1 \left(\sum |\alpha_k|^2 \right)^{1/2} \leq \left\| \sum \alpha_k \varphi(x - k) \right\|_2 \leq C_2 \left(\sum |\alpha_k|^2 \right)^{1/2}.$$

To simplify the notation, we have denoted the vector $(k_1, k_2) \in \mathbb{Z}^2$ by k and similarly $x = (x_1, x_2) \in \mathbb{R}^2$.

If this is the case, let V_0 denote the closed linear subspace of $L^2(\mathbb{R}^2)$ generated by the functions $\varphi(x - k)$, $k \in \mathbb{Z}^2$. Then it is easy to verify that the conditions in (4.18) hold and that $V_j \subset V_{j+1}$ when the V_j are defined by (4.19).

The pyramid algorithms associated with multiresolution analyses are the only ones that we will study in the following sections. They have some quite remarkable properties. For example, the restriction operator R_j is then an isomorphism between V_j and $l^2(\Gamma_j)$. In this case, the equation $R_{j-1} = T_j R_j$ can be solved directly; it is sufficient, in fact, to restrict the two sides of the space in V_j in order to invert R_j.

Not all pyramid algorithms are related to a multiresolution analysis. A counterexample is given by one of the pyramid algorithms that we presented in §4.3. In this example the function $\varphi(x)$ is $1/2$ on $[-1, 1]$ and 0 elsewhere. Thus

$$(4.24) \quad \|\varphi(x) - \varphi(x - 1) + \varphi(x - 2) + \cdots + (-1)^N \varphi(x - N)\|_2 = \frac{1}{\sqrt{2}},$$

whereas, according to (4.23), it should be of the order of $\sqrt{N}$.

4.6.   The orthogonal pyramids and wavelets.

Shortly after the discovery of quadrature mirror filters by Esteban and Galand, Woods and O'Neil [13] had the idea to apply this technique to image processing. They thus obtained the first example of an orthogonal pyramid. We are going to set aside, for the moment, the specific construction carried out by Woods and O'Neil using separable filters. Instead we will present the notion of an orthogonal pyramid in complete generality. Then we will return to the particular case where the quadrature mirror filters appear in the construction of an orthogonal pyramid.

The Burt and Adelson algorithm is flawed because it replaces information coded on N^2 pixels by new information whose description requires $\frac{4}{3}N^2$ pixels. This criticism, which we will analyze in a moment, is not justified because in many examples of real images, most of the values of the gray levels on the $\frac{4}{3}N^2$ pixels are in fact zero, and the unfavorable pixel count, where N^2 becomes $\frac{4}{3}N^2$, occurs very rarely.

Let us examine, however, why the information has been wasted or, more precisely, where the inefficient coding occurs. At the start, the image f has been coded on N^2 pixels. Next, we replace this by the couple $[T_0(f), (1 - T_0^*T_0)(f)]$, which is composed of the coding for the trend and the complete description of the fluctuations around the trend. The description of $T_0(f)$ requires $\frac{1}{4}N^2$ pixels, whereas the description of $f - T_0^*T_0(f)$ continues to require N^2 pixels. In all, we use $N^2 + \frac{1}{4}N^2$ pixels. At the next step, the pixel count becomes $N^2 + \frac{1}{4}N^2 + \frac{1}{16}N^2$, and so on... At the end, we will have used $N^2 + \frac{1}{4}N^2 + \frac{1}{16}N^2 + \cdots + 1$, or approximately $\frac{4}{3}N^2$ pixels. The "wasted" pixels appear because the fluctuations $f - T_j^*T_j(f)$ have not been coded efficiently.

The orthogonal pyramids are a particular class of pyramid algorithms that code the fluctuations with $\frac{3}{4}N^2$ pixels. With this scheme there is no waste. When the original image f is replaced by coding the trend and the fluctuation, the required pixels are $\frac{1}{4}N^2$ and $\frac{3}{4}N^2$, respectively, and the volume of data remains constant.

We say that a pyramid algorithm is orthogonal if, in the sense of the usual scalar product of $l^2(\Gamma_0)$, the trend $T_0^* T_0(f)$ and the fluctuations $f - T_0^* T_0(f)$ around this trend are orthogonal for each image f defined on the grid Γ_0.

Write $H = l^2(\Gamma_0)$, $H_0 = T_0^* T_0(H)$, and $H_1 = (I - T_0^* T_0)H$. If the pyramid algorithm is orthogonal, then $H = H_0 \oplus H_1$. Since the dimension of H_0 is a quarter that of H, the dimension of H_1 is $3/4 \dim H$, as announced.

An equivalent definition of "orthogonal pyramids" requires the adjoint T_0^* of the operator T_0 to be a partial isometry. (Recall that T_0^* is defined on $l^2(\Gamma_{-1})$ with values in $l^2(\Gamma_0)$.) This takes us back to one of the characteristic properties, in dimension one, of the "low-pass filter" T_0 in a pair of quadrature mirror filters, (T_0, T_1). And this observation prompts us, in dimension two, to look for the corresponding second filter, T_1. We will see in a moment that three filters are necessary in this case.

But before this, we show how to construct some pyramid algorithms. We return to the "transfer function" $m_0(\xi, \eta)$ defined by (4.11). The pyramid algorithm is orthogonal if and only if

(4.25)
$$|m_0(\xi, \eta)|^2 + |m_0(\xi + \pi, \eta)|^2 + |m_0(\xi, \eta + \pi)|^2$$
$$+ |m_0(\xi + \pi, \eta + \pi)|^2 = 1.$$

This condition is completely analogous to the one on the transfer function $m_0(\xi)$ in the case of two quadrature mirror filters (Chapter 3, §3.3).

Continuing this comparison, we consider the function $\varphi(x, y)$ in $L^2(\mathbb{R}^2) \cap L^1(\mathbb{R}^2)$ defined by (4.12) and normalized by $\iint \varphi(x, y) dx \, dy = 1$. We might expect that the sequence $\varphi(x - k, y - l)$, $k, l \in \mathbb{Z}$, is orthonormal, and this is true in most cases. However, the proof involves a delicate limit process, passing from the discrete to the continuous, and certain orthogonal pyramids do not lead to orthogonal sequences.

This difficulty already appeared in dimension one for the quadrature mirror filters. The condition we assume here on $m_0(\xi, \eta)$, which is sufficient to allow passage from the "discrete to the continuous," is completely analogous to the condition we used in dimension one.

It is sufficient to assume that $m_0(\xi, \eta) \neq 0$ if $-\frac{\pi}{2} \leq \xi \leq \frac{\pi}{2}$ and $-\frac{\pi}{2} \leq \eta \leq \frac{\pi}{2}$. Then $\varphi(x - k, y - l)$, $k, l \in \mathbb{Z}$, is an orthonormal basis of a closed subspace V_0 of $L^2(\mathbb{R}^2)$. At the same time, $2^j \varphi(2^j x - k, 2^j y - l)$, $k, l \in \mathbb{Z}^2$, is an orthonormal basis for the subspace V_j; the extension operator $P_j : l^2(\Gamma_j) \to V_j$ is an isometric isomorphism; and the restriction operator $R_j : L^2(\mathbb{R}^2) \to l^2(\Gamma_j)$ is decomposed into the orthogonal projection operator from $L^2(\mathbb{R}^2)$ onto V_j followed by the inverse isomorphism $P_j^{-1} : V_j \to l^2(\Gamma_j)$.

This allows us to define explicitly the transition operators $T_j : l^2(\Gamma_j) \to l^2(\Gamma_{j-1})$, which, we recall, were defined implicitly by $T_j R_j = R_{j-1}$. Use the operator P_j to identify $l^2(\Gamma_j)$ with V_j, and similarly use P_{j-1} to identify $l^2(\Gamma_{j-1})$ with V_{j-1}. Having made these identifications, the transition operator $T_j : l^2(\Gamma_j) \to l^2(\Gamma_{j-1})$ corresponds quite simply to the orthogonal projection of V_j on V_{j-1}, which is $P_{j-1} T_j P_j^{-1}$ in our notation.

We next define W_j to be the orthogonal complement of V_j in V_{j+1}. Thus we have $V_{j+1} = V_j \oplus W_j$. It is easy to verify—using once again the "isometric interpretations" given by $P_j : l^2(\Gamma_j) \to V_j$ and $P_{j+1} : l^2(\Gamma_{j+1}) \to V_{j+1}$—that this orthogonal decomposition corresponds precisely to the orthogonal decomposition of a function into its trend and fluctuation, and this latter decomposition is the definition of orthogonal pyramids.

We come now to the two-dimensional generalization of the quadrature mirror filters. In dimension two, we consider four operators T_0, R_1, R_2, and R_3. All four are defined on $l^2(\mathbb{Z}^2)$ with values in $l^2(2\mathbb{Z}^2)$. We ask that these four operators commute with the even translations $\tau \in 2\mathbb{Z}^2$ and that

$$(4.26) \qquad \|f\|^2 = \|T_0(f)\|^2 + \|R_1(f)\|^2 + \|R_2(f)\|^2 + \|R_3(f)\|^2,$$

for all f belonging to $l^2(\mathbb{Z}^2)$. The left term is of course computed in $l^2(\mathbb{Z}^2)$, whereas each term on the right is computed in $l^2(2\mathbb{Z}^2)$.

One of the important results in the theory of orthogonal pyramids is the existence and constructability of these operators R_1, R_2, and R_3. Furthermore, if the impulse response $w(k, l)$ of T_0 decreases rapidly at infinity, the operators R_1, R_2, and R_3 can be constructed to have this same property.

Once R_1, R_2, and R_3 are constructed, we can construct the corresponding wavelets ψ_1, ψ_2, and ψ_3. Assuming that $m_0(\xi, \eta) \neq 0$ if $|\xi| \leq \frac{\pi}{2}$ and $|\eta| \leq \frac{\pi}{2}$, these three wavelets are defined by

$$(4.27) \qquad \psi_j(x, y) = 4 \sum_k \sum_l \omega_j(k, l) \varphi(2x + k, 2y + l), \qquad j = 1, 2, \text{ or } 3,$$

where $\omega_j(k, l)$ denotes the impulse response of R_j.

Thus, under quite general conditions, the orthogonal pyramids lead to orthonormal wavelet bases, and this development proceeds by way of the two-dimensional generalization of quadrature mirror filters.

We move on to the two-dimensional generalization of Mallat's algorithm. The "exact reconstruction" identity,

$$(4.28) \qquad I = T_0^* T_0 + R_1^* R_1 + R_2^* R_2 + R_3^* R_3,$$

is deduced immediately from (4.26). Identity (4.28) provides a particularly elegant solution to the problem of coding the fluctuation $f - T_0^* T_0(f)$. This fluctuation is exactly

$$R_1^* R_1(f) + R_2^* R_2(f) + R_3^* R_3(f).$$

The three operators R_1^*, R_2^*, and R_3^* are partial isometries, and this allows us to code the fluctuation $f - T_0^* T_0(f)$ with the three sequences $R_1(f)$, $R_2(f)$, $R_3(f)$. These three sequences belong to $l^2(2\Gamma_0)$ when $f \in l^2(\Gamma_0)$, and thus the coding of each of them uses only one pixel in four. Hence, three-fourths of the pixels are used to code the fluctuation, whereas one-fourth is used to code the trend. Consequently there are no longer any wasted pixels.

We can now return to the algorithm and give it a much more precise formulation. This is illustrated by the following scheme. The horizontal arrows correspond to coding the trends, whereas the oblique arrows represent coding the three fluctuations.

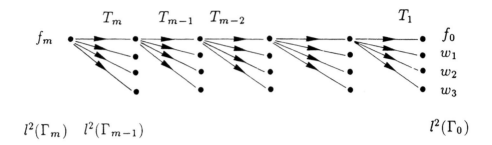

$$l^2(\Gamma_m) \quad l^2(\Gamma_{m-1}) \qquad\qquad\qquad\qquad\qquad l^2(\Gamma_0)$$

The wavelets appear in the asymptotic limit of this scheme. This limit is taken on the number of steps m, which must tend to infinity. We start with a fixed function $f(x, y)$ belonging to $L^2(\mathbb{R}^2)$. We "restrict" f to the fine grid Γ_m using the classic scheme. This means we have a fixed regular function g, which decreases rapidly at infinity and whose Fourier transform $\hat{g}(\xi, \eta)$ satisfies $\hat{g}(0, 0) = 1$ and $\hat{g}(2k\pi, 2l\pi) = 0$ if $(k, l) \neq (0, 0)$. We write $g_m(x, y) = 4^m g(2^m x, 2^m y)$. Finally, f_m is the restriction to the (fine) grid Γ_m of the (filtered) image $f * g_m$.

If we still assume that $m_0(\xi, \eta)$ does not vanish on $\left[-\frac{\pi}{2}, \frac{\pi}{2}\right] \times \left[-\frac{\pi}{2}, \frac{\pi}{2}\right]$ and that the pyramid is orthogonal as defined by (4.25), then Mallat's algorithm converges as the number of steps tends to infinity. The limit of this process is another algorithm, namely, the decomposition of the original image $f(x, y)$ in the orthonormal basis composed of the following four sequences: $\varphi(x - k, y - l)$, $2^j \psi_1(2^j x - k, 2^j y - l)$, $2^j \psi_2(2^j x - k, 2^j y - l)$, and $2^j \psi_3(2^j x - k, 2^j y - l)$, where $k, l \in \mathbb{Z}$ and $j \in \mathbb{N}$.

This means that if we fix the index j of the grid Γ_j, and if we examine the "out-puts" of Mallat's algorithm that are defined on this grid, then the limits of their coefficients are, respectively,

$$2^j \int f(x, y)\varphi(2^j x - k, 2^j y - l)dx\,dy,$$

$$2^j \int f(x, y)\psi_1(2^j x - k, 2^j y - l)dx\,dy,$$

$$2^j \int f(x, y)\psi_2(2^j x - k, 2^j y - l)dx\,dy,$$

$$2^j \int f(x, y)\psi_3(2^j x - k, 2^j y - l)dx\,dy.$$

Albert Cohen established this result under very general hypotheses [4]; of these, the most convenient is that $m_0(\xi, \eta) \neq 0$ if $|\xi| \leq \frac{\pi}{2}$ and $|\eta| \leq \frac{\pi}{2}$.

The beauty of this theory leads one to think that it provides the correct response to the image-processing problem. *Indeed, the image is decomposed by wavelet analysis into information that is independent from one scale to another, and this agrees with the general philosophy expressed in the introduction. These independent packets of information are represented by the trend in V_0 and the fluctuations $f_j \in W_j$, whose orthogonal sum is equal to f. The characteristic scale of W_j is 2^j. Furthermore, each $f_j \in W_j$ is itself decomposed into orthogonal* *components according to the basis* $2^j \psi(2^j x - k, 2^j y - l)$, $(k, l) \in \mathbb{Z}^2$, $\psi = \psi_1, \psi_2$, *or* ψ_3.

The Haar system is the simplest example of orthogonal wavelets, and it has been used for image processing for a long time. However, it has the disadvantage that, following quantization, it introduces rather harsh "edge effects," thus producing unpleasant images.

This prompts us to say a few words about the quantization problem. From the point of view of the mathematician, all orthonormal bases allow the signal to be reconstructed exactly. This is not the point of view of the numerical analyst or image specialist. In practice, the coefficients from the decomposition must be quantized, whether we like it or not. These approximations arise from the machine accuracy or are imposed by a desire to compress the data. If it is true that

$$f(x, y) = \sum_{j \in J} \alpha_j e_j(x, y),$$

what happens to the first term when the α_j are replaced by coefficients $\tilde{\alpha}_j$ satisfying $|\tilde{\alpha}_j - \alpha_j| \leq \varepsilon$, where $\varepsilon > 0$ is related to the machine precision? Anything can happen if we are not using a robust orthonormal basis, and we know today that many orthogonal wavelet bases are infinitely more robust than the trigonometric basis.

In spite of this, orthogonal wavelets (and the corresponding pyramid algorithms) have not completely satisfied the experts in image processing. The criticized defect is the lack of symmetry. The function $\varphi(x)$ ought to be even, while the function $\psi(x)$ ought to be symmetric in the sense that $\psi(1 - x) = \psi(x)$. These properties are satisfied by certain orthogonal wavelets, but they do not hold for wavelets with compact support. The Haar system is the only exception.

This lack of symmetry is reflected in visible defects, again following quantization. These visible defects do not appear when one uses symmetric bi-orthogonal wavelets having compact support. We introduce these wavelets in the next section.

4.7.   Bi-orthogonal wavelets.

Following the pioneering work of Philippe Tchamitchian [11], Albert Cohen, Ingrid Daubechies, and Jean-Christophe Feauveau studied a remarkable generalization of the notion of orthonormal wavelet bases, namely, bi-orthogonal systems of wavelets. We begin with the one-dimensional case.

In place of an orthonormal basis of the form $2^{j/2} \psi(2^j x - k)$, $j, k \in \mathbb{Z}$, we use two Riesz bases, each the dual of the other, denoted by $\psi_{j,k}$ and $\tilde{\psi}_{j,k}$. The first is used for synthesis, and the second is used for analysis. This means that for all $f(x)$ belonging to $L^2(\mathbb{R})$

$$(4.29) \qquad\qquad f(x) = \sum_{-\infty}^{\infty} \sum_{-\infty}^{\infty} \alpha_{j,k} \psi_{j,k}(x),$$

where $\|f\|_2$ and $\left(\sum_{-\infty}^{\infty} \sum_{-\infty}^{\infty} |\alpha_{j,k}|^2 \right)^{1/2}$ are equivalent norms on $L^2(\mathbb{R})$ and where the coefficients are defined by

(4.30)
$$\alpha_{j,k} = \int_{-\infty}^{\infty} f(x)\overline{\tilde{\psi}_{j,k}(x)}\,dx.$$

Here
$$\psi_{j,k}(x) = 2^{j/2}\psi(2^j x - k), \quad \text{and} \quad \tilde{\psi}_{j,k}(x) = 2^{j/2}\tilde{\psi}(2^j x - k).$$

Up to this point we have only weakened the definition of the orthonormal wavelet bases. We have gained nothing. But now we are going to make considerably stronger demands by requiring $\psi(x)$ to be an essentially explicit function. For example, we can require that $\psi(x)$ be the following function:

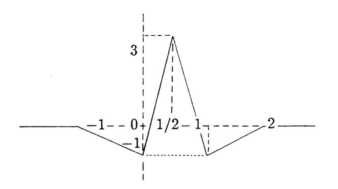

Then the general theory of Cohen, Daubechies, and Feauveau tells us that, for this choice of $\psi(x)$, $2^{j/2}\psi(2^j x - k)$, $j, k \in \mathbb{Z}$, is a Riesz basis for $L^2(\mathbb{R})$ and the dual basis has the same structure, which is given by $2^{j/2}\tilde{\psi}(2^j x - k)$, $j, k \in \mathbb{Z}$. However, in this particular case, the dual wavelet is not a continuous function. To repair this we need to take a more general approach.

We assume that ψ belongs to a set of functions that are continuous, have compact support, are linear on each interval $[k/2, (k+1)/2]$, $k \in \mathbb{Z}$, and are symmetric with respect to $1/2$ (in the sense that $\psi(1-x) = \psi(x)$). Then the Cohen–Daubechies–Feauveau theory states that we can choose a ψ from this set so that the dual wavelet $\tilde{\psi}$ is a function in the class C^r and has compact support.

Here is how ψ and $\tilde{\psi}$ are constructed. Start with the triangle function $\varphi(x) = \sup(1 - |x|, 0)$, which was mentioned in §4.2. Write $m_0(\xi) = (\cos \xi/2)^2$; then, by construction, $\hat{\varphi}(\xi) = m_0(\xi/2)\hat{\varphi}(\xi/2)$. Next consider $g_N(\xi) = c_N \int_\xi^\pi (\sin t)^{2N+1} dt$, where $c_N > 0$ is adjusted so that $g_N(0) = 1$.

If $\tilde{m}_0(\xi)$ is defined by $m_0(\xi)\tilde{m}_0(\xi) = g_N(\xi)$, then

(4.31)
$$m_0(\xi)\tilde{m}_0(\xi) + m_0(\xi + \pi)\tilde{m}_0(\xi + \pi) = 1.$$

(In the construction of the Daubechies wavelets, one imposed the condition $|m_0(\xi)|^2 = g_N(\xi)$.)

Define $\tilde{\varphi} \in L^2(\mathbb{R})$ by its Fourier transform

(4.32)
$$\hat{\tilde{\varphi}}(\xi) = \tilde{m}_0(\xi/2)\tilde{m}_0(\xi/4)\dots$$

Identity (4.31) is equivalent to

$$(4.33) \qquad \int \tilde{\varphi}(x)\varphi(x-k)dx = 0 \quad \text{if} \quad k \neq 0, \quad \text{and} \quad = 1 \quad \text{if} \quad k = 0.$$

The function $\tilde{\varphi}(x)$ is even, its support is the interval $[-2N, 2N]$, and $\tilde{\varphi}(x)$ is in the Hölder space C^r for all sufficiently large N.

It is clear that

$$\left\| \sum_{-\infty}^{\infty} \alpha_k \tilde{\varphi}(x-k) \right\|_2 \leq C \left(\sum_{-\infty}^{\infty} |\alpha_k|^2 \right)^{1/2},$$

but (4.33) implies that the inverse inequality also holds.

Thus one can consider the closed subspace $\tilde{V}_0 \subset L^2(\mathbb{R})$ for which $\tilde{\varphi}(x-k)$, $k \in \mathbb{Z}$, is a Riesz basis. If the subspaces $\tilde{V}_j$ are defined by (4.19), this sequence forms a multiresolution analysis of $L^2(\mathbb{R})$.

In the same way, let V_0 be the closed subspace of $L^2(\mathbb{R})$ for which $\varphi(x-k)$, $k \in \mathbb{Z}$, is a Riesz basis and construct the V_j again by (4.19).

The two multiresolutions (V_j) and $(\tilde{V}_j)$ are the duals of each other. This duality is used to define the subspaces W_j and $\tilde{W}_j$: f belongs to W_j if f belongs to V_{j+1} and if $\int_{-\infty}^{\infty} f(x)\overline{u(x)}dx = 0$ for all $u \in \tilde{V}_j$.

The wavelets ψ and $\tilde{\psi}$ will be constructed so that $\psi(x-k)$, $k \in \mathbb{Z}$, is a Riesz basis for W_0 and, similarly, $\tilde{\varphi}(x-k)$, $k \in \mathbb{Z}$, is a Riesz basis for $\tilde{W}_0$. For this, we define

$$m_1(\xi) = e^{-i\xi}\tilde{m}_0(\xi+\pi) \quad \text{and} \quad \tilde{m}_1(\xi) = e^{-i\xi}m_0(\xi+\pi)$$

and define the Fourier transforms $\hat{\psi}$ and $\hat{\tilde{\psi}}$ of ψ and $\tilde{\psi}$ by

$$\hat{\psi}(\xi) = m_1(\xi/2)\hat{\varphi}(\xi/2), \qquad \hat{\tilde{\psi}}(\xi) = \tilde{m}_1(\xi/2)\hat{\tilde{\varphi}}(\xi/2).$$

We write $\psi_{j,k}(x) = 2^{j/2}\psi(2^j x - k)$ and define $\tilde{\psi}_{j,k}(x)$ similarly.

The only properties that are not clear are that the family $\psi_{j,k}, j, k \in \mathbb{Z}$, is a Riesz basis for $L^2(\mathbb{R})$ and that the same is true for the $\tilde{\psi}_{j,k}$. These Riesz bases are the duals of each other. Furthermore, the function $\psi(x)$ is as simple as it is explicit: it is continuous; it has compact support; it is linear on each interval $[k/2, (k+1)/2]$; and the values $\psi(k/2)$ are explicit rational numbers. Finally, we have $\psi(1-x) = \psi(x)$, and the symmetry, which Daubechies's wavelets lack, is reestablished.

In dimension two, we use the wavelets $\varphi(x)\psi(y)$, $\varphi(y)\psi(x)$, and $\psi(x)\psi(y)$, as in the orthogonal case. Then the dual wavelets are $\tilde{\varphi}(x)\tilde{\psi}(y)$, $\tilde{\varphi}(y)\tilde{\varphi}(x)$, and $\tilde{\psi}(x)\tilde{\psi}(y)$.

By using these bi-orthogonal wavelets and an efficient method of vector quantization for coding the wavelet coefficients, Barlaud [1] has achieved a compression ratio of order 100.

Bibliography

[1] M. ANTONINI, M. BARLAUD, I. DAUBECHIES, AND P. MATHIEU, *Image coding using vector quantization in the wavelet transform domain*, IEEE International Conference on Acoustics, Speech, and Signal Processing, Albuquerque, NM, April 1990, pp. 2297–2300.

[2] E. H. ADELSON, R. HINGORANI, AND E. SIMONCELLI, *Orthogonal pyramid transforms for image coding*, SPIE, Visual Communications and Image Processing II, Cambridge, MA, October 27–29, Vol. 845, pp. 50–58.

[3] P. J. BURT AND E. H. ADELSON, *The Laplacian pyramid as a compact image code*, IEEE Trans. Comm., COM-31 (1983), pp. 532–540.

[4] A. COHEN, *Ondelettes, analysis multirésolutions et traitement numérique du signal*, Thesis, CEREMADE, University of Paris-Dauphine, Paris, France, September 1990.

[5] J. C. FEAUVEAU, *Analyse multirésolution par ondelettes non orthogonales et bases de filtres numérique*, Thesis, University of Paris-South, Paris, France, January 1990.

[6] J. FROMENT, *Traitement d'images et applications de la transformée en ondelettes*, Thesis, CEREMADE, University of Paris-Dauphine, Paris, France, November 1990.

[7] J. FROMENT AND J. M. MOREL, *Analyse multiéchelle, vision stéréo et ondelettes*, in Les ondelettes en 1989, P. G. Lemarié, ed., Lecture Notes in Mathematics 1438, Springer-Verlag, Berlin, New York, 1990, pp. 51–80.

[8] S. MALLAT, *A theory for multiresolution signal decomposition: The wavelet representation*, IEEE Trans. Pattern Anal. Machine Intelligence, 11 (1989), pp. 674–693.

[9] D. MARR, *Vision, A computational investigation into the human representation and processing of visual information*, W. H. Freeman and Co., New York, 1982.

[10] Y. MEYER, *Ondelettes*, Hermann, Paris, 1990.

[11] P. TCHAMITCHIAN, *Biorthogonalité et Théorie des Opérateurs*, Revista Matemática Iberoamericana, 3 (1987), pp. 163–189.

[12] M. VETTERLI, *Wavelets and filter banks: Relationships and new results*, IEEE Conference on Acoustics, Speech and Signal Processing, Albuquerque, NM, April 1990, pp. 1723–1726.

[13] J. W. WOODS AND S. O'NEIL, *Subband coding of images*, IEEE Trans. Acous. Speech Signal Process., 34 (1986), pp. 1278–1288.

CHAPTER 5

Time-Frequency Analysis for Signal Processing

5.1. Introduction.

Dennis Gabor (1946) and Jean Ville (1947) both addressed the problem of developing a mixed signal representation in terms of a double sequence of *elementary signals*, each of which occupies a certain domain in the *time-frequency plane*. In the following sections we will define what is meant by "time-frequency plane" and "mixed representation," and we will suggest several choices for "elementary signals."

Roger Balian [1] tackled the same problem and expressed the motivation for his work in these terms:

> One is interested, in communication theory, in representing an oscillating signal as a superposition of *elementary wavelets,* each of which has a rather well defined *frequency* and *position in time.* Indeed, useful information is often conveyed by both the emitted frequencies and the signal's temporal structure (music is a typical example). The representation of a signal as a function of time provides a poor indication of the spectrum of frequencies in play, while, on the other hand, its Fourier analysis masks the point of emission and the duration of each of the signal's elements. An appropriate representation ought to combine the advantages of these two complementary descriptions; at the same time, it should be discrete so that it is better adapted to communication theory.

Similar criticism of the usual Fourier analysis, as applied to acoustic signals, is found in the celebrated work of Ville:

> If we consider a passage [of music] containing several measures (which is the least that is needed) and if a note, *la* for example, appears once in the passage, harmonic analysis will give us the corresponding frequency with a certain amplitude and a certain phase, without localizing the *la* in time. But it is obvious that there are moments during the passage when one does not hear the *la*. The [Fourier] representation is nevertheless mathematically correct because the phases of the notes near the *la* are arranged so as to destroy this note through interference when it is not heard and to reinforce it, also through interference, when it is heard; but if there is in this idea a cleverness that speaks well for mathematical analysis, one must not ignore the fact that it is also a distortion of reality: indeed when the *la* is not heard, the true reason is that the *la* is not emitted.

Thus it is desirable to look for a mixed definition of a signal of the sort advocated by Gabor: at each instance, a certain number of frequencies are present, giving volume and timbre to the sound as it is heard; each frequency is associated with a certain partition of time that defines the intervals during which the corresponding note is emitted. One is thus led to define an instantaneous spectrum as a function of time, which describes the structure of the signal at a given instant; the spectrum of the signal (in the usual sense of the term), which gives the frequency structure of the signal based on its total duration, is then obtained by putting together all of the instantaneous spectrums in a precise way by integrating them with respect to time. In a similar way, one is led to a distribution of frequencies with respect to time; by integrating these distributions, one reconstructs the signal. . .

Ville thus proposed to unfold the signal in the time-frequency plane in such a way that this development would lead to a mixed representation in time-frequency atoms. The choice of these time-frequency atoms would be guided by an energy distribution of the signal in the time-frequency plane.

The time-frequency atoms proposed by Gabor are constructed from the function $g(t) = \pi^{-1/4} \exp(-t^2/2)$ and are defined by

$$(5.1) \qquad\qquad w(t) = h^{-1/2} \exp(i\omega t) g((t - t_0)/h).$$

The parameters ω and t_0 are arbitrary real numbers, whereas h is positive. The meaning of these three parameters is the following: ω is the average frequency of $w(t)$, $h > 0$ is the duration of $w(t)$, and $t_0 - h$, $t_0 + h$ are the start and finish of the "note" $w(t)$. Naturally, all this depends on the convention used to define the "pass band" of $g(t)$.

The essential problem is to describe an algorithm that allows a given signal to be decomposed, in an optimal way, as a linear combination of judiciously chosen time-frequency atoms.

The set of all time-frequency atoms (with ω and t_0 varying arbitrarily in the time-frequency plane and $h > 0$ covering the whole scale axis) is a collection of elementary signals that is much too large to provide a unique representation of a signal as a linear combination of time-frequency atoms. Each signal admits an infinite number of representations, and this leads us to choose the best among them according to some criterion. This criterion might be the one suggested by Ville: The decomposition of a signal in time-frequency atoms is related to a synthesis, and this synthesis ought logically to be done in accordance with an analysis. The analysis proposed by Ville will be described in the following sections. However, Ville did not explain how the results of the analysis would lead to an effective synthesis.

A similar program (the definition of time-frequency atoms, analysis, and synthesis) was proposed by Jean-Sylvan Liénard [3]:

We consider the speech signal to be composed of elementary waveforms, wf, (windowed sinusoids), each one defined by a small number of parameters. A waveform model (wfm) is a sinusoidal signal multiplied by a windowing function. It is not to be confused with the signal segment, wf, that it is supposed to approximate. Its total duration can be decomposed into attack (before the max-

imum of the envelope), and decay. In order to minimize spectral ripples, the envelope should present no 1st or 2nd order discontinuity. The initial discontinuity is removed through the use of an attack function (raised sinusoids) such that the total envelope is null at the origin, and maximum after a short time. Although exponential damping is natural in the physical world, we choose to model the decaying part of the wfs with another raised sinusoid. Actually we see the wf as a perceptual unit, and not necessarily as the response of a format filter to a voicing impulse...

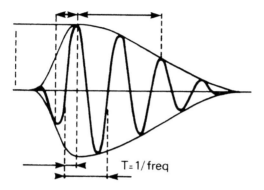

Liénard's "time-frequency atoms" are thus different from those used by Gabor. They are, however, based on analogous principles. We have $w(t) = A(t)\cos(\omega t + \varphi)$, where ω represents the average frequency of the emitted "note" and where the envelope $A(t)$ incorporates the attack and decay. The principal difference is that, in the "atoms" of Liénard, the duration of the attack and that of the decay are independent. Thus Liénard's "atoms" depend on four independent parameters, and the optimal representation of a speech signal as a linear combination of time-frequency atoms is more difficult to obtain. Some empirical methods exist, and they lead to wonderful results for synthesizing the singing voice. I had the chance to hear the Queen of the Night's grand aria from Mozart's *Magic Flute* interpreted by "time-frequency atoms." This was not a copy of the human voice; it involved the creation of a purely numerical (superhuman) voice. This was commissioned by Pierre Boulez, the director of the Institut de Recherches Coordonnées Acoustique-Musique.

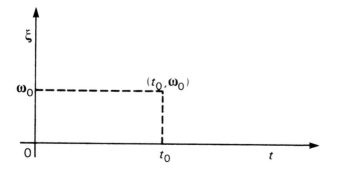

5.2.   The time-frequency plane.

The time-frequency plane serves the acoustician the way music paper serves the musician. To continue this metaphor, the signal analysis that we seek to effect should be compared to an exercise called a "musical dictation," which consists in writing down the notes on hearing a passage of music.

5.3.   The Wigner–Ville transform.

We begin by presenting the point of view of Ville. We will then indicate how to interpret the results in terms of the theory of pseudodifferential operators as expressed in Hermann Weyl's formalism. This will bring us back to work done by the physicist Eugene P. Wigner (1932).

Ville, searching for an "instantaneous spectrum," wanted to display the energy of a signal in the time-frequency plane and to obtain an *energy density* $W(t, \xi)$ having (at least) the following properties:

$$(5.2) \qquad \int_{-\infty}^{\infty} W(t, \xi) \frac{d\xi}{2\pi} = |f(t)|^2,$$

$$(5.3) \qquad \int_{-\infty}^{\infty} W(t, \xi) dt = |\hat{f}(\xi)|^2,$$

where $\hat{f}(\xi)$ denotes the Fourier transform of f. These two properties reflect the program that we presented in the introduction: at each instant t, the function $W(t, \xi)$ gives an instantaneous Fourier analysis of the signal $f(t)$, and (5.2) is the Plancherel formula. The same remark holds for (5.3): $|\hat{f}(\xi)|^2$ comes from the contributions of all instants t, and one hopes that $|\hat{f}(\xi)|^2$ is more precisely analyzed by using these "individual contributions."

Properties (5.2) and (5.3) are clearly not sufficient to define $W(t, \xi) = W_f(t, \xi)$. We impose two other conditions, namely, "Moyal's formula"

$$(5.4) \qquad \iint W_f(t, \xi) W_g(t, \xi) dt \frac{d\xi}{2\pi} = \left| \int f(t) \overline{g(t)} dt \right|^2,$$

which plays the role of Parseval's identity, and the requirement that if

$$(5.5) \qquad f(t) = h^{-1/2} \exp(i\omega t) g((t - t_0)/h)$$

then

$$(5.6) \qquad W_f(t, \xi) = 2 \exp\left(-\frac{(t - t_0)^2}{h^2}\right) \exp(-h^2(\xi - \omega)^2).$$

Let's stop a moment to examine (5.6). The second member is a function of (t, ξ) that is localized on the rectangle of the time-frequency plane defined by $|t - t_0| \le h$, $|\xi - \omega| \le \frac{1}{h}$. This localization corresponds exactly to the frequency content of the time-frequency atom $f(t)$. Up to a normalization factor, the second member of (5.6) is the solution to the localization problem in the time-frequency plane that we want for our time-frequency atom.

We now come to the general definition of the Wigner–Ville transform of a signal $f(t)$. First assume that the energy $\int_{-\infty}^{\infty} |f(t)|^2 dt$ is finite. Then define

$$(5.7) \qquad W(t, \xi) = \int_{-\infty}^{\infty} f\left(t + \frac{\tau}{2}\right) \overline{f}\left(t - \frac{\tau}{2}\right) e^{-i\xi\tau} d\tau.$$

We notice immediately that $W(t, \xi)$ is real and continuous in both variables. It is easy to verify that $W(t, \xi)$ has the properties indicated in (5.2) to (5.6).

If $f(t) = f(-t)$, then $W(0, 0) = 2 \int_{-\infty}^{\infty} |f(\tau)|^2 d\tau > W(t, \xi)$ for all other pairs (t, ξ).

5.4.   The computation of certain Wigner–Ville transforms.

We begin by treating the case of signals with finite energy. If $W(t, \xi)$ is the Wigner–Ville transform of $f(t)$, then

$$W(t, \xi - \omega) \quad \text{is the transform of} \quad e^{i\omega t} f(t),$$
$$W(t - t_0, \xi) \quad \text{is the transform of} \quad f(t - t_0),$$

and

$$W\left(\frac{t}{a}, a\xi\right) \quad \text{is the transform of} \quad \frac{1}{a} f\left(\frac{t}{a}\right), a > 0.$$

Knowing that the transform of $g(t) = \pi^{-1/4} \exp(-t^2/2)$ is $2 \exp(-t^2 - \xi^2)$, we deduce immediately that the Wigner–Ville transform of

$$\frac{1}{\sqrt{h}} e^{i\omega t} g\left(\frac{t - t_0}{h}\right)$$

is

$$2 \exp\left(-\frac{(t - t_0)^2}{h^2} - h^2(\xi - \omega)^2\right).$$

Here are some other useful observations. The Wigner–Ville transformation of a function characterizes the function up to multiplication by a constant of modulus 1. We will prove this in the next section when we establish the connection between the Wigner–Ville transform and the pseudodifferential calculus. The Wigner–Ville transform of $f(-t)$ is $W(-t, -\xi)$ when the transform of $f(t)$ is $W(t, \xi)$. Multiplying $f(t)$ by a real or complex constant λ results in the transform $W(t, \xi)$ being multiplied by $|\lambda|^2$. Thus we need to consider only the case where $\int_{-\infty}^{\infty} |f(t)|^2 dt = 1$ when we are working with signals of finite energy.

Not all functions $W(t, \xi)$ of the two variables t and ξ are the Wigner–Ville transform of some signal $f(t)$.

We consider a positive-definite quadratic form

$$Q(t,\xi) = p\xi^2 + 2r\xi t + qt^2 \quad (p > 0, q > 0, pq > r^2)$$

and ask when $2\exp(-Q(t,\xi))$ is the Wigner–Ville transform of a signal $f(t)$. In view of the preceding remarks, we must have $f(-t) = f(t)$, and we assume that $\int_{-\infty}^{\infty}|f(t)|^2 dt = 1$. But then $\iint W(t,\xi)dt\,d\xi = 2\pi$, which implies that $pq - r^2 = 1$. We will show that this necessary condition is also sufficient for $2\exp(-Q(t,\xi))$ to be the Wigner–Ville transform of a signal.

For this, we observe that if $W(t,\xi)$ is the Wigner–Ville transform of $f(t)$, then the transform of $f(t)e^{i\alpha t^2/2}$, where α is a real constant, is $W(t,\xi - \alpha t)$.

Our quadratic form $p\xi^2 + 2r\xi t + qt^2$ can also be written as $Q(t,\xi) = p(\xi - (r/p)t)^2 + (t^2/p)$ since $pq - r^2 = 1$. The Wigner–Ville transform of $p^{-1/4}g(t/\sqrt{p})$, where $g(t) = \pi^{-1/4}e^{-t^2/2}$, is $2\exp(-(t^2/p) - p\xi^2)$, and thus that of $p^{-1/4}g(t/\sqrt{p})\exp(i(r/2p)t^2)$ is $2\exp(-Q(t,\xi))$. Our problem is solved.

More generally $2\exp(-Q(t - t_0, \xi - w))$ is the Wigner–Ville transform of the signal

(5.8) $$f(t) = p^{-1/4}g\left(\frac{t - t_0}{\sqrt{p}}\right)\exp\left(i\frac{r}{2p}(t - t_0)^2\right)\exp(iwt),$$

which is called a "chirp." The quadratic form $Q(t,\xi) = p\xi^2 + 2r\xi t + qt^2$ is subject to the condition $r^2 - pq = -1$.

Here is an important identity involving the Wigner–Ville transform. We have

(5.9) $$\frac{1}{2\pi}\int_{-\infty}^{\infty}\hat{f}(\omega + \xi/2)\overline{\hat{f}}(\omega - \xi/2)e^{i\xi t}d\xi = \int_{-\infty}^{\infty}f(t + \tau/2)\overline{f}(t - \tau/2)e^{-i\omega\tau}d\tau.$$

In other words, if $W(t,\omega)$ is the Wigner–Ville transform of $f(t)$, then $W(\omega, -t)$ is that of $\frac{1}{\sqrt{2\pi}}\hat{f}(\xi)$. Here again $\hat{f}$ denotes the Fourier transform of f.

Another very useful fact is that the Wigner–Ville transform of an arbitrary function is always real, but it is not always positive. This second remark is the source of a great deal of difficulty in the interpretation of $W(t,\xi)$.

Ville interpreted the Wigner–Ville transform $W(t,\xi)$ of a normalized signal $f(t)$ as a probability density in the time-frequency plane. If this probability density were concentrated in several well-delimited rectangles in the time-frequency plane, this would lead to a decomposition of the signal in terms of the corresponding time-frequency atoms.

This program has not led to an effective algorithm. The reason for this failure is that if $f(t)$ is the sum of two time-frequency atoms $e^{i\omega_1 t}g(t-t_1)+e^{i\omega_2 t}g(t-t_2)$, then

$$W(t,\xi) = W_1(t,\xi) + W_2(t,\xi) + W_3(t,\xi) + W_4(t,\xi),$$

where the terms $W_1(t,\xi)$ and $W_2(t,\xi)$ are the "square" terms already calculated and where W_3 and W_4 are two "cross" terms. But these cross terms do not tend to zero if $\omega_2 - \omega_1$ or if $t_2 - t_1$ tends to infinity. In fact,

$$|W_3(t,\xi)| = |W_4(t,\xi)| = 2\exp(-(t-t_3)^2 - (\xi - \omega_3)^2)$$

where $t_3 = \frac{1}{2}(t_1+t_2)$ and $\omega_3 = \frac{1}{2}(\omega_1+\omega_2)$. These "cross" terms are thus artifacts that are localized in the time-frequency plane midway between the corresponding square terms.

The fact that the Wigner–Ville transform is not, in general, positive and the fact that its localization in the time-frequency plane does not necessarily imply the presence of time-frequency atoms are two independent properties. This can be seen by considering the signal $f(t) = e^{-t}$ for $t \geq 0$, and $f(t) = 0$ elsewhere. Then $W(t,\xi) = e^{-2t}\frac{\sin 2t\xi}{\xi}$ if $t \geq 0$ and $W(t,\xi) = 0$ otherwise.

There exists, however, a simple way to make the Wigner–Ville transform positive. It suffices to smooth it appropriately. Indeed, if f_1 and f_2 are two arbitrary functions (with finite energy), then

$$(5.10) \quad \iint W_{f_1}(t-u, \xi-v)W_{f_2}(u,v)du\,dv = 2\pi \left| \int_{-\infty}^{\infty} f_1(s)\overline{f_2(t-s)}e^{-i\xi s}ds \right|^2.$$

If, in particular, $f_2(t) = \frac{1}{\sqrt{h}}g\left(\frac{t}{h}\right)$, where $g(t)$ is the normalized Gaussian and where $h > 0$ is arbitrary, then we have $W_{f_2}(u,v) = 2\exp(-(u^2/h^2) - h^2 v^2)$ and the smoothing function is a Gaussian kernel. The mean value one obtains is the square of the modulus of the scalar product of f_1 and the time-frequency atom $\frac{1}{\sqrt{h}}g\left(\frac{s-t}{h}\right)e^{i\xi s}$, centered at t, with width h and average frequency ξ.

But the Wigner–Ville transform $W(t,\xi)$ of a signal f can also be smoothed by using a kernel of the form $\frac{1}{\pi}\exp(-Q(t,\xi))$, where $Q(t,\xi)$ is one of the quadratic forms previously studied. One obtains a positive contribution that is the square of the modulus of the scalar product of f with a "chirp."

5.5.   The Wigner–Ville transform and pseudodifferential calculus.

The following considerations allow us to relate the Wigner–Ville transform to quantum mechanics and the work of Wigner. We are going to forget signal-processing problems for the moment and go directly to dimension n. The analogue of the time-frequency plane is the phase space $\mathbb{R}^n \times \mathbb{R}^n$ whose elements are pairs (x,ξ), where x is a position and ξ is a frequency.

We start with a "symbol" $\sigma(x,\xi)$ defined on phase space. Certain technical hypotheses have to be made about this symbol to ensure convergence of the following integral when f belongs to a reasonable class of test functions, and we will deal with this in a moment.

Following the formalism of Weyl, we associate with the symbol $\sigma(x,\xi)$ the pseudodifferential operator $\sigma(x,D)$ defined by

$$(5.11) \quad (2\pi)^n \sigma(x,D)[f](x) = \iint \sigma\left(\frac{x+y}{2},\xi\right)e^{i(x-y)\cdot\xi}f(y)dy\,d\xi,$$

where the integral is over $\mathbb{R}^n \times \mathbb{R}^n$. Define the kernel $K(x,y)$ associated with the symbol $\sigma(x, \xi)$ by

$$(2\pi)^n K(x, y) = \int \sigma\left(\frac{x+y}{2}, \xi\right) e^{i(x-y)\cdot\xi} d\xi$$

(5.12)

$$= (2\pi)^n L\left(\frac{x+y}{2}, x-y\right).$$

This says that the symbol $\sigma(x, \xi)$ is the partial Fourier transform, in the variable u, of the function $L(x, u)$ and that the kernel $K(x, y)$ that interests us is $L\left(\frac{x+y}{2}, x-y\right)$. We can also write, in the inverse sense, $L(x, y) = K\left(x + \frac{y}{2}, x - \frac{y}{2}\right)$, and this allows us to recover the symbol $\sigma(x, \xi)$ by writing

(5.13) $$\sigma(x, \xi) = \int K\left(x + \frac{y}{2}, x - \frac{y}{2}\right) e^{-iy\cdot\xi} dy.$$

Thus we are led to hypotheses about the symbols that are the reflections, through the partial Fourier transform, of hypotheses that we may wish to make about the kernels. If we admit all the kernel distributions $K(x, y)$ belonging to $S'(\mathbb{R}^n \times \mathbb{R}^n)$, then there will be no restrictions on $\sigma(x, \xi)$ other than the condition that

$$\sigma(x, \xi) \in S'(\mathbb{R}^n \times \mathbb{R}^n).$$

An immediate consequence of (5.13) is this: If $\sigma(x, \xi)$ is the symbol for the operator T, then $\overline{\sigma(x, \xi)}$ is the symbol for the adjoint operator T^*.

Finally, we consider a function f belonging to $L^2(\mathbb{R}^n)$ and satisfying $\|f\|_2 = 1$. Let P_f denote the orthogonal projection operator that maps $L^2(\mathbb{R}^n)$ onto the linear span of f. Then the kernel $K(x, y)$ of P_f is $f(x)\overline{f(y)}$ and the corresponding Weyl symbol is

(5.14) $$\sigma(x, \xi) = \int f\left(x + \frac{y}{2}\right) \overline{f}\left(x - \frac{y}{2}\right) e^{-iy\cdot\xi} dy.$$

Returning to dimension one, we have the following result: *The Wigner–Ville transform of the function f is the Weyl symbol of the orthogonal projection operator onto that function f.* From this it is clear that the Wigner–Ville transform of f characterizes f, up to multiplication by a constant of modulus 1.

5.6.   The Wigner–Ville transform and instantaneous frequency.

In Ville's fundamental work (which has essentially been the source for this chapter), he makes a careful distinction between the instantaneous frequency of a signal (assumed to be real) and the instantaneous spectrum of frequencies given by the Wigner–Ville transform.

More precisely, let $f(t)$ be a real signal with finite energy. Ville writes $f(t) = \text{Re } F(t)$, where $F(t)$ is the corresponding analytic signal: $F(t)$ is the restriction to the real axis of a function $F(z)$ that is holomorphic in the upper half-plane $\text{Im } z > 0$ and belongs to the Hardy space $H^2(\mathbb{R})$.

Ville then writes $F(t) = A(t)e^{i\varphi(t)}$, where $A(t)$ is the modulus of $F(t)$ and $\varphi(t)$ is its argument. He defines the instantaneous frequency of $f(t)$ by $\frac{d}{dt}\varphi(t)$.

This definition requires the function $f(t)$ to have additional regularity properties. Otherwise $\varphi(t)$ could be as irregular as an arbitrary bounded, measurable function, and the instantaneous frequency would then be a very singular object. This also raises a problem about the continuity of $\varphi(t)$ so as not to introduce Dirac measures in $\frac{d}{dt}\varphi(t)$.

We will not deal with these difficulties, and we assume that Ville's formal definitions make sense. This, of course, clearly limits the class of analyzed signals.

Following Ville, we define the instantaneous spectrum of $f(t)$ as the Wigner–Ville transform $W(t,\xi)$ of the analytic signal $F(t)$.

An easy calculation shows that, for all real or complex-valued functions $u(t)$, one has

$$\int_{-\infty}^{\infty}\int_{-\infty}^{\infty}\xi u(t+\tau/2)\overline{u}(t-\tau/2)e^{-i\tau\xi}d\xi\,d\tau = -\pi i(u'(t)\overline{u}(t) - u(t)\overline{u}'(t)).$$

Applying this identity to $u(t) = F(t)$, it becomes

$$\frac{1}{2\pi}\int_{-\infty}^{\infty}\xi W(t,\xi)d\xi = \varphi'(t)|F(t)|^2 = \varphi'(t)\left(\frac{1}{2\pi}\int_{-\infty}^{\infty}W(t,\xi)d\xi\right).$$

If $W(t,\xi)$ is positive or zero, then $\frac{1}{2\pi}W(t,\xi)$ will be a probability density (when $\int_{-\infty}^{\infty}|F(t)|^2dt = 1$) and the *instantaneous frequency will be the average of the frequency ξ computed with respect to the instantaneous spectrum.*

Similarly, we can try to compute the analogue of the variance of the variable ξ with respect to the instantaneous spectrum. This is $\int_{-\infty}^{\infty}(\xi - \varphi'(t))^2W(t,\xi)d\xi$. The calculation is completely general and does not rely on the assumption that $F(t) = A(t)e^{i\varphi(t)}$ is an analytic function. We obtain

$$(5.15)\qquad \int_{-\infty}^{\infty}(\xi - \varphi'(t))^2W(t,\xi)d\xi = -\pi\left[\frac{d^2}{dt^2}(\log A(t))\right]A^2(t).$$

If, in particular, $A(t) = 1$, the second member is zero, and $W(t,\xi)$ cannot be positive or zero unless it is concentrated on the curve $\xi = \varphi'(t)$, which represents the graph of the instantaneous frequency. Since $2\pi A^2(t) = \int_{-\infty}^{\infty}W(t,\xi)d\xi$, the "variance" of ξ is equal to $-\frac{1}{2}(d^2/dt^2)(\log A(t))$.

Here are two examples of the calculation of the instantaneous frequency. Suppose first that the original signal $f(t)$ is real, equal to 1 on the interval $[-T,T]$, and 0 outside this interval. The corresponding analytic signal is then

$$F(t) = f(t) + \frac{i}{\pi}\log\left|\frac{t+T}{t-T}\right|.$$

The phase $\varphi(t)$ of $F(t)$ is continuous on the whole real line, odd, equal to $\frac{\pi}{2}$ if $t \geq T$ (and thus $-\frac{\pi}{2}$ if $t \leq -T$), and strictly increasing on $[-T,T]$. The instantaneous frequency is 0 outside the interval $[-T,T]$, strictly positive on $(-T,T)$, equal to $\frac{2}{\pi T}$ at 0, and increases from $\frac{2}{\pi T}$ to $+\infty$ as t traverses the interval $[0,T)$.

As one could have guessed, *the instantaneous Fourier analysis proposed by Ville is not even a local property. This means that knowing the signal in an arbitrary large interval centered at t_0 is not sufficient to calculate the instantaneous frequency at t_0*. The operation responsible for this anomaly is the calculation of the analytic signal $F(t)$ associated with $f(t)$; as everyone knows, the kernel of the Hilbert transform decreases slowly at infinity.

This discussion shows that the signals to which the Ville theory applies are necessarily academic signals (whose algorithmic structure does not change over time) or asymptotic signals whose behavior on a short time interval is equivalent to that of a normal signal over a much longer duration.

The second example of a calculation of the instantaneous frequency is for the signal $f(t) = \cos t^2$, which is a chirp of infinite duration. The calculation of the corresponding analytic signal is interesting because it exhibits two different asymptotic behaviors depending on whether t tends to $+\infty$ or $-\infty$. Indeed, the Fourier transform of this analytic signal $F(t)$ is 0 for $\xi < 0$ and is equal to $\frac{1}{2}(\sqrt{\pi i}\ e^{-i\xi^2/4} + \sqrt{-\pi i}\ e^{i\xi^2/4})$ for $\xi > 0$. It follows that $F(t)$ is asymptotically equal to e^{it^2} when t tends to $+\infty$ and to e^{-it^2} when t tends to $-\infty$. The instantaneous frequency of $\cos t^2$ is thus equal to $2t + \varepsilon(t)$ when t tends to $+\infty$ and to $-2t + \varepsilon(t)$ when t tends to $-\infty$. In both cases, $\varepsilon(t)$ tends to 0 when $|t|$ tends to $+\infty$.

5.7. The Wigner–Ville transform of asymptotic signals.

As we have already seen in §5.5, the Wigner–Ville transform can be generalized to the case where, instead of being a signal of finite energy, $f(t)$ is an arbitrary tempered distribution. We limit our discussion to three examples where $f(t) = e^{i\varphi(t)}$. We begin with the particular case where $\varphi(t) = \omega t$. ($f(t)$ is an analytic signal only when $\omega \geq 0$, but the calculations that follow do not depend on this type of hypothesis.)

The Wigner–Ville transform of $f(t) = e^{i\omega t}$ is $2\pi\delta_0(\xi - \omega)$, where $\delta_0(\xi)$ is the Dirac measure at the origin. Then the instantaneous frequency given by the Wigner–Ville transform is simply ω.

Next, if $f(t) = e^{i\alpha t^2/2}$, for α real, the Wigner–Ville transform of $f(t)$ is $2\pi\delta_0(\xi - \alpha t)$. This is a distribution (in fact, a measure) supported by the line $\xi = \alpha t$. The corresponding "instantaneous frequency" is αt, and both members of (5.15) are 0 in this case. In fact, this statement is not correct because $e^{i\alpha t^2/2}$ is not an analytic signal. However, as we have already observed, $e^{i\alpha t^2/2}$ is asymptotic to an analytic signal when α is strictly positive and when t tends to $+\infty$.

Finally we come to the case where $f(t) = e^{i\alpha t^3}$ with $\alpha > 0$. The Wigner–Ville transform of this function is easily calculated and is

$$\int \exp[i(\alpha\tau^3/4 + 3\alpha\tau t^2 - \omega\tau)]d\tau = 2\pi\left(\frac{4}{3\alpha}\right)^{1/3} A\left(\left(\frac{4}{3\alpha}\right)^{1/3}(3\alpha t^2 - \omega)\right),$$

where

$$A(\omega) = \frac{1}{2\pi} \int_{-\infty}^{\infty} e^{i((s^3/3)+\omega s)} ds$$

is the Airy function. Here again the Wigner–Ville transform of the function $e^{i\alpha t^3}$ is "essentially" concentrated around the curve $\xi = 3\alpha t^2$, which is the graph of the instantaneous frequencies. All of this must be put in quotation marks since $e^{i\alpha t^3}$ is not an analytic signal. But the signal is asymptotically analytic because the Airy function decreases exponentially when ω tends to $+\infty$.

5.8. Return to the problem of optimal decomposition in time-frequency atoms.

As we indicated in the introduction to this chapter, the analysis of the energy distribution of a signal in the time-frequency plane was, for Ville, a precondition for his search for optimal decompositions in time-frequency atoms.

Consider the example of the signal $f(t) = e^{i\alpha t^2/2}$. In the time-frequency plane, its energy is concentrated on the line $\xi = \alpha t$. If we try to decompose this function in time-frequency atoms of the kind advocated by Gabor, this comes down to covering the line $\xi = \alpha t$ in an optimal way with "Heisenberg boxes" of area 1. We can think of these squares as leading to Gabor wavelets of the form $e^{i\alpha k t} g(t - k)$. Finally, all of this leads to approximating $f(t) = e^{i\alpha t^2/2}$ with the sum of the series $\sum_{-\infty}^{\infty} e^{-i\alpha k^2/2} e^{i\alpha k t} g(t - k)$, and this is a poor approximation.

The other major shortcoming of the Wigner–Ville transform is that it is not always positive. One might have thought, in light of the example of the signal $f(t) = e^{i\omega_1 t} g(t - t_1) + e^{i\omega_2 t} g(t - t_2)$, that the places where the Wigner–Ville transform is positive correspond to the time-frequency atoms and that the places where it oscillates were simply artifacts. But this is not the case, as we see from the example of the signal $f(t) = e^{-t}$ for $t \geq 0$; $f(t) = 0$ otherwise.

We are forced to conclude that the Wigner–Ville transform yields only imperfect information about the distribution of energy in the time-frequency plane. There does not exist an algorithm that allows us to find an atomic decomposition of a signal by using the Wigner–Ville transform.

Bibliography

[1] R. BALIAN, *Un principe d'incertitude fort en théorie du signal ou en mécanique quantique*, C. R. Acad. Sci. Paris, Sér. II, 292 (1981), pp. 1357–1361.

[2] P. FLANDRIN, *Some aspects of non-stationary signal processing with emphasis on time-frequency and time-scale methods*, in Wavelets, J. M. Combe, A. Grossman, and Ph. Tchamitchian, eds., Springer-Verlag, Berlin, 1989, pp. 68–98.

[3] J. S. LIÉNARD, *Speech analysis and reconstruction using short-time, elementary waveforms*, ICASSP 87.

[4] J. VILLE, *Théorie et applications de la notion de signal analytique*, C&T, Laboratoire de Télécommunications de la Société Alsacienne de Construction Mécanique, 2éme A. No. 1 (1948).

CHAPTER 6

Time-Frequency Algorithms Using Malvar Wavelets

6.1. Introduction.

This chapter is the logical sequel of the preceding one. We will introduce algorithms that allow us to decompose a given signal $s(t)$ into a linear combination of time-frequency atoms. The time-frequency atoms that we use are denoted by $f_R(t)$ and are coded by the Heisenberg rectangles R (with sides parallel to the axes and with area 1 or 2π depending on the normalization). If $R = [a, b] \times [\alpha, \beta]$, we require that the function f_R be essentially supported on the interval $[a, b]$ and that its Fourier transform $\hat{f}_R$ be essentially supported on $[\alpha, \beta]$. We also ask that the algorithmic structure of $f_R(t)$ be simple and explicit to facilitate numerical processing in real time. The decomposition

$$(6.1) \qquad s(t) = \sum_0^\infty \alpha_j f_{R_j}(t)$$

cannot be unique, and we take advantage of this flexibility by looking for optimal decompositions, which here means they contain the fewest possible terms.

The point of view of Ville (and of numerous other signal-processing experts) is that it is first necessary to understand the physics of the process and that "the algorithms will follow." Unfortunately, algorithms associated with the Wigner–Ville transform have never followed, they have never existed, and the Wigner–Ville transform is an analytic technique that does not lead to a synthesis or to the transmission of information.

Our approach here is to favor synthesis and transmission over analysis. The "time-frequency atoms" that we use are completely explicit: They are either Malvar wavelets or wavelet packets, and we will immediately write down the "atomic decompositions" of the type (6.1). This means that the synthesis will be direct, whereas the analysis will consist in choosing—with the use of an entropy criterion—the most effective synthesis, which is the one that leads to optimal compression. Thus the analysis proceeds according to algorithmic criteria and not according to physics, and it is not at all clear that this approach leads to a signal analysis that reveals physical properties having a real meaning. For example, Marie Farge had the idea to apply the algorithm to images of simulated two-dimensional turbulence. The algorithm extracted, in order, coherent structures, vorticity filaments, and so on down the scale. This is amazing because the algorithm is not based on an analysis that takes into consideration the underlying physics of fully developed turbulence.

After these general remarks, it is time to specify the algorithms. There are two options: Malvar wavelets and wavelet packets. With the first option, the signal is segmented adaptively and optimally, and then the segments are analyzed using classical Fourier analysis. The second option, "wavelet packets," reverses the order of these operations and first filters the signal adaptively; the analysis in the time variable is then imposed by the algorithm.

Ville (1947) proposed two types of analysis. He wrote: "We can either: first cut the signal into slices (in time) with a switch; then pass these different slices through a system of filters to analyze them. Or we can: first filter different frequency bands; then cut these bands into slices (in time) to study their energy variations." The first approach leads us to "Malvar wavelets" and the second to "wavelet packets."

6.2. Malvar wavelets: A historical perspective.

The discovery of *Malvar wavelets* (Henrique Malvar, 1987) falls within the general framework of windowed Fourier analysis. The window is denoted by $w(t)$, and it allows the signal $s(t)$ to be cut into "slices" that are regularly spaced in time $w(t-bl)s(t)$, $l = 0, \pm1, \pm2, \dots$ The parameter $b > 0$ is the average length of these "slices." Next, following Ville, one does a Fourier analysis on these slices, which reduces to calculating the coefficients $\int e^{-iakt}w(t-bl)s(t)dt$, where $a > 0$ must be related to b and where $k = 0, \pm1, \pm2, \dots$ This is thus the same as taking the scalar products of the signal $s(t)$ with the "wavelets"

$$w_{k,l}(t) = e^{iakt}w(t - bl).$$

This analysis technique was proposed by Gabor (1946), in which case the $w(t)$ was the Gaussian. The "Gabor wavelets" lead to serious algorithmic difficulties and, more generally, Low and Balian showed in the early 1980s that if $w(t)$ is sufficiently regular and well localized, then the functions $w_{k,l}$, $k, l \in \mathbb{Z}$, can never be an orthonormal basis for $L^2(\mathbb{R})$. More precisely, if the two integrals $\int_{-\infty}^{\infty}(1 + |t|)^2|w(t)|^2dt$ and $\int_{-\infty}^{\infty}(1 + |\xi|)^2|\hat{w}(\xi)|^2d\xi$ are both finite, the functions $w_{k,l}$, $k, l \in \mathbb{Z}$, cannot be an orthonormal basis of $L^2(\mathbb{R})$.

The crude window defined by $w(t) = 1$ on the interval $[0, 2\pi)$ and $w(t) = 0$ elsewhere escapes this criterion. By choosing $a = 1$ and $b = 2\pi$, the windowed analysis consists in restraining the signal to each interval $[2l\pi, 2(l+1)\pi)$ and using Fourier series to analyze each of the corresponding functions. But the functions obtained by this crude segmentation are not 2π-periodic, and the Fourier analysis will highlight this lack of periodicity and interpret it as a discontinuity or an abrupt variation in the signal.

One way to attenuate these numerical artifacts—without, however, eliminating them—is to use the Discrete Cosine Transform (DCT). We describe the continuous version of this transform.

On each interval $[2l\pi, 2(l+1)\pi)$, we analyze the signal $s(t)$ using the orthonormal basis composed of the functions $\frac{1}{\sqrt{2\pi}}$ and $\frac{1}{\sqrt{\pi}}\cos\left(\frac{k}{2}t\right)$, $k = 1, 2, 3, \dots$ If $s(t)$

is a very regular function, the numerical artifacts produced by the segmentation are reduced from an order of magnitude $1/k$ to order $1/k^2$.

The physicist Kenneth Wilson was the first to have the idea that one could get around the problem presented in the Balian–Low theorem by imitating the DCT and using a segmentation created with very regular windows. Wilson alternated the DCT with the Discrete Sine Transform (DST) according to whether l is even or odd; l denotes the position of the interval, and the DST uses the orthonormal basis of functions $\frac{1}{\sqrt{\pi}} \sin\left(\frac{k}{2}t\right)$, $k = 1, 2, 3, \ldots$.

Wilson's construction has been the point of departure for numerous efforts, the most notable of which is due to Ingrid Daubechies, Stéphane Jaffard, and Jean-Lin Journé. They used a window $w(t)$ having the property that both it and its Fourier transform decay exponentially, and they constructed $w(t)$ so that the functions $u_{k,l}$, $k = 1, 2, 3, \ldots, l \in \mathbb{Z}$, and $u_{0,l}$, $l \in 2\mathbb{Z}$, defined by

$$(6.2) \qquad u_{k,l}(t) = \sqrt{2}\, w(t - 2l\pi) \cos\left(\frac{k}{2}t\right), \qquad l \in 2\mathbb{Z}, \qquad k = 1, 2, \ldots$$

$$(6.3) \qquad u_{0,l}(t) = w(t - 2l\pi), \qquad\qquad\qquad l \in 2\mathbb{Z}, \qquad k = 0,$$

and

$$(6.4) \qquad u_{k,l}(t) = \sqrt{2}\, w(t - 2l\pi) \sin\left(\frac{k}{2}t\right), \qquad l \in 2\mathbb{Z} + 1, \qquad k = 1, 2, \ldots$$

constitute an orthonormal basis for $L^2(\mathbb{R})$.

Malvar did not know about this work. He discovered a family of orthonormal bases $u_{k,l}(t)$, which have exactly the same algorithmic structure described by (6.2), (6.3), and (6.4), but where the choice of the window $w(t)$ is simpler and more explicit. In fact, Malvar had only these hypotheses:

$$(6.5) \qquad\qquad w(t) = 0 \quad \text{if} \quad t \leq -\pi \quad \text{or} \quad t \geq 3\pi$$

$$(6.6) \qquad\qquad 0 \leq w(t) \leq 1 \quad \text{and} \quad w(2\pi - t) = w(t)$$

$$(6.7) \qquad\qquad w^2(t) + w^2(-t) = 1 \quad \text{if} \quad -\pi \leq t \leq \pi.$$

Then the construction is the same, and the sequence $u_{k,l}$ defined by (6.2), (6.3), and (6.4) is an orthonormal basis for $L^2(\mathbb{R})$. In Malvar's construction, the window $w(t)$ can be very regular (infinitely differentiable, for example), but the Fourier transform of w cannot have exponential decay. Condition (6.5) prevents it, and this condition plays an essential role in the demonstrations.

Although similar, the solutions found by Daubechies, Jaffard, and Journé, on the one hand, and by Malvar, on the other, are not linked by a logical connection.

6.3.   Windows with variable lengths.

Coifman and the author tackled the problem of modifying the preceding constructions to obtain windows with variable lengths that could be defined arbitrarily. The construction by Daubechies, Jaffard, and Journé does not extend to this context, while that of Malvar generalizes, without the slightest difficulty, to the case of arbitrary windows.

We begin with an arbitrary partition of the real line into adjacent intervals $[a_j, a_{j+1}]$, where $\ldots < a_{-1} < a_0 < a_1 < a_2 < \ldots$, $\lim_{j \to +\infty} a_j = +\infty$, and $\lim_{j \to -\infty} a_j = -\infty$. Write $l_j = a_{j+1} - a_j$ and let $\alpha_j > 0$ be positive numbers that are small enough so that $l_j \geq \alpha_j + \alpha_{j+1}$ for all $j \in \mathbb{Z}$.

The windows $w_j(t)$ that we use will essentially be the characteristic functions of the intervals $[a_j, a_{j+1}]$; the role played by the disjoint intervals $(a_j - \alpha_j, a_j + \alpha_j)$ is to allow the windows to overlap, which is necessary if we want the windows to be regular.

More precisely, we impose the following conditions:

(6.8) $0 \leq w_j(t) \leq 1$ for all $t \in \mathbb{R}$,

(6.9) $w_j(t) = 1$ if $a_j + \alpha_j \leq t \leq a_{j+1} - \alpha_{j+1}$,

(6.10) $w_j(t) = 0$ if $t \leq a_j - \alpha_j$ or $t \geq a_{j+1} + \alpha_{j+1}$,

(6.11) $w_j^2(a_j + \tau) + w_j^2(a_j - \tau) = 1$ if $|\tau| \leq \alpha_j$,

(6.12) $w_{j-1}(a_j + \tau) = w_j(a_j - \tau)$ if $|\tau| \leq \alpha_j$.

Clearly, these conditions allow the windows $w_j(t)$ to be infinitely differentiable.

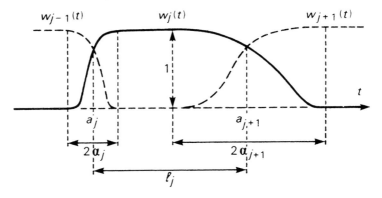

It is clear that $\sum_{-\infty}^{\infty} (w_j(t))^2 = 1$, identically on the whole real line.

Finally, we come to the Malvar wavelets. They appear in two distinct forms. The first is given by

(6.13) $u_{j,k}(t) = \sqrt{\dfrac{2}{l_j}} w_j(t) \cos \left[\dfrac{\pi}{l_j} \left(k + \dfrac{1}{2} \right) (t - a_j) \right],$

with $k = 0, 1, 2, \ldots$ and $j \in \mathbb{Z}$.

The second form consists in alternating the cosines and sines according to whether j is even or odd. Thus we have three distinct expressions for the second form:

(6.14) $u_{j,k}(t) = \sqrt{\dfrac{2}{l_j}} w_j(t) \cos \dfrac{k\pi}{l_j} (t - a_j)$

if $j \in 2\mathbb{Z}$ and $k = 1, 2, \ldots$

$$(6.15) \qquad u_{j,k}(t) = \sqrt{\frac{1}{l_j}} w_j(t) \quad \text{if} \quad j \in 2\mathbb{Z} \quad \text{and} \quad k = 0,$$

and

$$(6.16) \qquad u_{j,k}(t) = \sqrt{\frac{2}{l_j}} w_j(t) \sin \frac{k\pi}{l_j}(t - a_j)$$

if $j \in 2\mathbb{Z} + 1$ and $k = 1, 2, \ldots$

The functions $u_{j,k}(t)$, $j \in \mathbb{Z}$, $k = 0, 1, 2, \ldots$ given by (6.13) are an orthonormal basis for $L^2(\mathbb{R})$, and the same is true for the functions defined by (6.14), (6.15), and (6.16).

Here are two graphs of Malvar wavelets:

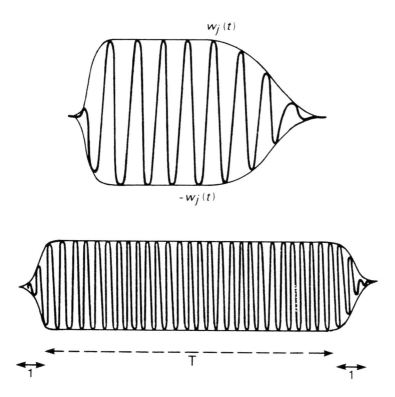

Note the extreme similarity between the Malvar wavelets and the time-frequency atoms proposed by Liénard. In particular, the Malvar wavelets are constructed with an attack (whose duration is $2\alpha_j$), a stationary period (which lasts $l_j - \alpha_j - \alpha_{j+1}$), and then a decay (which lasts $2\alpha_{j+1}$). The ability to choose, arbitrarily and independently, the duration of the attack, then that of the stationary section, and finally the duration of the relaxation is precisely what differentiates the Malvar wavelets from the preceding constructions (Gabor or Daubechies–Jaffard–Journé).

Of course, it is important to make good use of the freedom-of-choice at our disposal. We will see how to do this in the following sections.

6.4. Malvar wavelets and time-scale wavelets.

In 1985, Pierre-Gilles Lemarié and I constructed a function $\psi(t)$ belonging to the Schwartz class $S(\mathbb{R})$ such that $2^{j/2}\psi(2^j t - k)$, $j, k \in \mathbb{Z}$, is an orthonormal basis for $L^2(\mathbb{R})$. In addition, the Fourier transform of ψ is zero outside the intervals $\left[-\frac{8\pi}{3}, -\frac{2\pi}{3}\right]$ and $\left[\frac{2\pi}{3}, \frac{8\pi}{3}\right]$. We will see that these wavelets $2^{j/2}\psi(2^j t - k)$, $j, k \in \mathbb{Z}$, constitute a particular case of the general Malvar construction.

This is quite surprising because the Lemarié–Meyer wavelets constitute a "time-scale" algorithm, whereas the Malvar wavelets are a "time-frequency" algorithm. There is thus an incompatibility.

In fact, it is by analyzing the Fourier transform $\hat{f}$ of a (arbitrary) function f in an appropriate Malvar basis that we arrive at the analysis by Lemarié–Meyer wavelets.

We begin with the following observation: The Malvar wavelets allow us to analyze functions defined on a half-line, and this is contrary to what can be done with the Daubechies–Jaffard–Journé wavelets. The segmentation of $(0, \infty)$ that we use is the "natural" division into dyadic intervals $[2^j, 2^{j+1}]$, $j \in \mathbb{Z}$. Then it is natural to choose the windows $w_j(x)$, associated with these intervals, to be of the form $w_j(x) = w(2^{-j}x)$. Thus the whole construction rests on the precise choice for the function $w(x)$. For this, we make the following choices in accordance with conditions (6.8)–(6.12): $w(x)$ is zero outside the interval $[2/3, 8/3]$, $w(2x) = w(2 - x)$ for $2/3 \le x \le 4/3$, and $w^2(x) + w^2(2 - x) = 1$ on the same interval. Then $a_j = 2^j$, $\alpha_j = \frac{1}{3}2^j$, and $l_j = 2^j = \alpha_j + \alpha_{j+1}$.

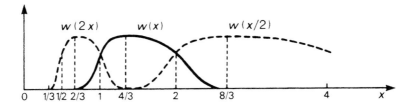

The Malvar wavelets of type (6.13) are then

(6.17) $$u_{j,k}(x) = \sqrt{2}\,2^{-j/2}\cos[\pi(k + 1/2)2^{-j}x]w(2^{-j}x).$$

We can just as easily replace all of the cosines in (6.13) by sines and thus obtain a second orthonormal basis for $L^2[0, \infty]$ of the form

(6.18) $$v_{j,k}(x) = \sqrt{2}\,2^{-j/2}\sin[\pi(k + 1/2)2^{-j}x]w(2^{-j}x).$$

We next extend $w(x)$ to the whole real line by making it an even function: $w(-x) = w(x)$. This gives a natural even extension for the functions $u_{j,k}$ and an odd extension for the $v_{j,k}$.

Finally, the complete collection of extended functions

(6.19) $$\left\{ \frac{1}{\sqrt{2}}u_{j,k}(x), \frac{1}{\sqrt{2}}v_{j,k}(x); \ j \in \mathbb{Z}, \ k = 0, 1, \ldots \right\}$$

is an orthonormal basis for $L^2(\mathbb{R})$.

It follows that the set of functions $\frac{1}{2}(u_{j,k} + iv_{j,k})$, $\frac{1}{2}(u_{j,k} - iv_{j,k})$ is also an orthonormal basis for $L^2(\mathbb{R})$.

Next, we observe that

(6.20) $$\frac{1}{2}(u_{j,k} + iv_{j,k})(x) = 2^{-(j+1)/2}w(2^{-j}x)\exp(i\pi(k+1/2)2^{-j}x)$$

and, that by letting $k^* = -1 - k$, we have

(6.21) $$\frac{1}{2}(u_{j,k} - iv_{j,k})(x) = 2^{-(j+1)/2}w(2^{-j}x)\exp(i\pi(k^*+1/2)2^{-j}x).$$

The conclusion is that the sequence

(6.22) $$2^{-(j+1)/2}w(2^{-j}x)\exp(i\pi(k+1/2)2^{-j}x), \qquad j, k \in \mathbb{Z},$$

is an orthonormal basis for $L^2(\mathbb{R})$.

Denote the Fourier transform of the function $\frac{1}{\sqrt{2}}w(x)e^{i\pi x/2}$ by $\psi(t)$. The function $\psi(t)$ is real and satisfies $\psi(\pi - t) = \psi(t)$. Then the sequence

(6.23) $$\frac{1}{\sqrt{2\pi}}2^{j/2}\psi(2^j t - k\pi), \qquad j, k \in \mathbb{Z},$$

is an orthonormal basis for $L^2(\mathbb{R})$. To regain the usual form, $2^{j/2}\psi(2^j t - k)$, $j, k \in \mathbb{Z}$, it suffices to replace t by πt.

It is clearly possible to require that $w(x)$ be an infinitely differentiable function, in which case $\psi(t)$ will be a function in the Schwartz class $\mathcal{S}(\mathbb{R})$.

We recall the program of Ville. There were two possible approaches: either segment the signal appropriately and follow this by Fourier analysis, or pass the signal through a bank of filters and then study the individual outputs of the filter banks.

Here we have selected the second approach. The filter bank that we use is defined by the transfer functions $w(2^{-j}x)$, where $w(x)$ is the even window used above.

6.5.   Adaptive segmentation and the split-and-merge algorithm.

We are not going to create a segmentation *ex nihilo*, but we will modify an existing segmentation to produce a new one. The modification operation is described in this section. A segmentation is modified by adjusting the partition (a_j) that defines the segmentation, and this is done by iterating the following elementary modifications: An elementary modification consists in suppressing a point a_j of the partition, which means that the two intervals $[a_{j-1}, a_j]$ and

$[a_j, a_{j+1}]$ are combined into a single interval, namely, $[a_{j-1}, a_{j+1}]$. The other intervals remain unchanged. This operation is called "merging." The inverse operation consists in adding an extra point α between the points a_j and a_{j+1}, which results in replacing the interval $[a_j, a_{j+1}]$ by the two intervals $[a_j, \alpha]$ and $[\alpha, a_{j+1}]$. This inverse operation is called "splitting."

A split-and-merge algorithm provides a criterion to decide when and where to use one or the other of these elementary operations.

We are going to examine the effect of these operations on a Malvar basis. We will show that an elementary operation induces an elementary modification of the basis that is very easy to calculate. The following remark is the point of departure for this discussion.

For each fixed j, let W_j denote the closed subspace of $L^2(\mathbb{R})$ generated by the functions $u_{j,k}(t)$, $k = 0, 1, 2, \ldots$ described by (6.13). Then $f(t)$ belongs to W_j if and only if $f(t) = w_j(t)q(t)$, where $q(t)$ belongs to $L^2[a_j - \alpha_j, a_{j+1} + \alpha_{j+1}]$ and satisfies the following two conditions: $q(a_j + \tau) = q(a_j - \tau)$ if $|\tau| \leq \alpha_j$ and $q(a_{j+1} + \tau) = -q(a_{j+1} - \tau)$ if $|\tau| \leq \alpha_{j+1}$. There are no conditions that need to be satisfied on the interval $[a_j + \alpha_j, a_{j+1} - \alpha_{j+1}]$.

To use the suggestive language of Ronald Coifman and Guido Weiss, W_j is a "dipole" with a positive polarity at a_j and a negative polarity at a_{j+1}. The intervals $(a_j - \alpha_j, a_j + \alpha_j)$ where the polarities interact are pairwise disjoint.

From here, the merging algorithm is trivial. Removing the point a_j of the partition amounts to replacing the two subspaces W_{j-1} and W_j by their direct orthogonal sum $W_{j-1} \oplus W_j$ without disturbing any of the other spaces $W_{j'}$, $j' \neq j-1$ and $j' \neq j$. But this, in turn, comes down to replacing the two windows w_{j-1} and w_j by the new window $\tilde{w}_j$ defined by $\tilde{w}_j(t) = (w_{j-1}^2(t) + w_j^2(t))^{1/2}$. The two lengths l_{j-1} and l_j are replaced by $\tilde{l}_j = l_{j-1} + l_j$, which changes the fundamental frequency in (6.13).

To fix our ideas, we start with a segmentation with intervals of length 1, $a_j = j$, and choose $w_j(t) = w(t - j)$ with $\alpha_j = 1/3$. Next, we look at which windows can appear as a result of the merging algorithm, which consists in removing intermediate points in the partition. The resulting windows and the corresponding wavelets will look like centipedes (see the figure of the second Malvar wavelet). The localization of these centipedes in the time-frequency plane is not optimal. This is because, in using the merging algorithm, we never change the values of the numbers α_j. In our example, we always keep $\alpha_j = 1/3$. This being the case, the algorithm allows us to replace only the partition consisting of all the integer intervals of length 1 by a partition that has arbitrary integer intervals $[a_j, a_{j+1}]$. The wavelets that appear are given by (6.13), and they are approximately localized in the time-frequency plane around the rectangles

$$R(j, k) = [a_j, a_{j+1}] \times \left[\frac{k\pi}{l_j}, \frac{(k+1)\pi}{l_j} \right], \qquad l_j = a_{j+1} - a_j.$$

We must now provide the criterion that allows us to decide when to use the dynamic split-and-merge algorithm. This means that we need to establish a numerical value to measure what is gained or lost by adding or deleting a point in the subdivision. This is the purpose of the next section.

6.6.   The entropy of a vector with respect to an orthonormal basis.

Let H denote a Hilbert space and let $(e_j)_{j \in J}$ be an orthonormal basis for H. Let x be a vector of H of length 1. We write $x = \sum_{j \in J} \alpha_j e_j$; the entropy of x relative to the basis e_j measures the number of significant terms in this decomposition. This entropy is defined by $\exp(-\sum_{j \in J} |\alpha_j|^2 \log |\alpha_j|^2)$.

If we have a collection $(e_j^\omega)_{j \in J}$ of orthonormal bases where ω ranges over a set Ω, we will choose for the analysis of x the particular basis (indexed by ω_0) that yields the minimum entropy.

This point of view poses three problems:

(6.24)   Does an optimal basis exist?

(6.25)   A diagnostic on the transmitted signal can be made
            impracticable by a compression algorithm whose
            only objective is efficiency.

(6.26)   The underlying energy criterion (the square
            of the norm in the Hilbert space H) can cause
            certain information in the signal to be given
            low priority, and this information can subsequently
            disappear in the compression, even though
            it may be crucial for the diagnostic.

In image analysis, all of the algorithms are based on calculations that are done from an energy function, which is defined as the quadratic mean value of the gray levels; the algorithm used to search for an optimal basis for compression does not escape this difficulty. The search for a Hilbert norm that is adapted exactly to the structure of the image is still an unsolved problem.

A concrete example of where these difficulties arise is a program to transmit radiographic images within large hospitals over the ordinary telephone lines. The compression needed for transmission and the quality of the received image that is required for the doctor to make a diagnosis are clearly antagonistic objectives. The interested reader can refer to [3], where this problem is discussed.

6.7.   The algorithm for finding the optimal Malvar basis.

We will examine in detail the particular case where the Hilbert space H is the space of signals $f(t)$ with finite energy, which is defined by $\int_{-\infty}^{\infty} |f(t)|^2 dt$. The quality of the compression will be measured only by this criterion.

The algorithm looks for "the best basis," which is the one that optimizes compression based on the reduction of transmitted data. The search is done by comparing the scores of a whole family of orthonormal bases of $L^2(\mathbb{R})$. These are Malvar bases, and they are obtained from segmentations of the real line into dyadic intervals. These intervals are systematically constructed in a scheme that moves from "fine" to "coarse." This means that we start with a segmentation by intervals of length 2^{-q}, where $q \geq 0$ is large enough to capture the finest details appearing in the signal.

The process consists in removing, if necessary, certain points in the segmentation and in replacing, at the same time, two contiguous dyadic intervals I' and I'' (appearing in the former segmentation) with the dyadic interval $I = I' \cup I''$. For example, $[2, 3]$ and $[3, 4]$ can become $[2, 4]$ with the disappearance of 3, but $[3, 4]$ and $[4, 5]$ will never become $[3, 5]$. For the point 4 to disappear, it would be necessary to wait for the eventual merging of the intervals $[0, 4]$ and $[4, 8]$.

By an obvious change of scale, we may assume that $q = 0$. The starting point is thus the segmentation where the "fine grid" is $\mathbb{Z}$. The intervals $[a_j, a_{j+1}]$ of §6.5 are now $[j, j + 1]$, and the first orthonormal basis to participate in the competition will be

$$(6.27) \qquad u_{j,k}(t) = \sqrt{2} \cos\left[\pi\left(k + \frac{1}{2}\right)(t - j)\right] w(t - j),$$

where $j \in \mathbb{Z}$, $k = 1, 2, 3, \ldots$

The other orthonormal bases that participate in the competition will all be obtained from this first one by merging. The regrouping algorithm that merges two orthonormal bases into one was described in detail in §6.5. We will limit its application to situations where $[a_{j-1}, a_j]$ and $[a_j, a_{j+1}]$ are the left (I') and (I'') right halves of a dyadic interval I.

Each partition of the real line into dyadic intervals of length greater than or equal to 1 thus defines canonically one of the orthonormal bases that are allowed to participate in the competition.

One reaches all of these partitions by iterating those elementary operations that combine the left and right halves of a dyadic interval and by traversing this tree structure, starting from the "fine grid."

We denote by $\mathcal{I}$ the collection of all the dyadic intervals I of length $|I| \geq 1$, and if $I = [a_j, a_{j+1}]$ is one of these dyadic intervals, we denote by w_I the window that was denoted by w_j in §6.5. In the same way, we denote by W_I the closed subspace of $L^2(\mathbb{R})$ that was denoted by W_j and by $w_I^{(k)}$ the orthonormal sequence defined by (6.13), which is now an orthonormal basis for W_I.

If I' and I'' are, respectively, the left and right halves of the dyadic interval I, then we have $W_I = W_{I'} \oplus W_{I''}$ and this direct sum is orthogonal.

The signal $f(t)$ that we wish to analyze optimally is normalized by $\int_{-\infty}^{\infty} |f(t)|^2 dt = 1$. To simplify the following discussion, we assume in addition that $f(t)$ is zero outside the interval $[1, T]$ for some sufficiently large T. Then f belongs to W_L if $L = [0, 2^l]$ and l is large enough.

It is easy to show that if m tends to $+\infty$, the entropy of f in the orthonormal basis $w_I^{(k)}$ of W_I, $I = [0, 2^m]$, also tends to infinity. Thus there exists some value of m after which $w_I^{(k)}$ no longer enters into the competition. In other words, the dyadic partitions that come into play will, in fact, be the partitions of $L = [0, 2^m]$ (for sufficiently large m) into dyadic intervals I of length $|I| \geq 1$. The number of partitions is thus finite, but it can be incredibly large (the order of magnitude being $2^{(2^m)}$). It remains to find a fast algorithm to search for the "best basis."

This is the algorithm that we are now going to describe. If I belongs to $\mathcal{I}$, then $c_I^{(k)} = \int_I f(t) w_I^{(k)}(t) dt$,

(6.28)
$$\varepsilon(I) = -\sum_0^\infty |c_I^{(k)}|^2 \log |c_I^{(k)}|^2,$$

and

(6.29)
$$\varepsilon^*(I) = \inf \sum_p \varepsilon(J_p),$$

where the lower bound is taken over all the partitions (J_p) of the interval I into dyadic intervals J_p belonging to $\mathcal{I}$. If $I = [j, j+1]$, then clearly $\varepsilon^*(I) = \varepsilon(I)$.

The problem that we must solve is thus clearly reduced to finding the optimal partition (J_p) when $I = L = [0, 2^m]$, the largest of the dyadic intervals involved in the competition. The calculation of $\varepsilon^*(L)$ and the determination of the optimal partition cannot be done directly because the number of cases to be considered is too large. We will calculate $\varepsilon^*(I)$ for $|I| = 2^n$ by induction on n. For $n = 0$, we must calculate $\varepsilon^*(I) = \varepsilon(I)$ for all intervals $I = [j, j+1]$ in $[0, 2^m]$. Next we proceed by induction on n, assuming that we have calculated $\varepsilon^*(I)$ for $|I| = 2^n$ and that we have determined the corresponding covering (J_p).

Suppose that $|I| = 2^{n+1}$ and let I' and I'' be the left and right halves of I. There are two cases:

(6.30)   If $\varepsilon(I) \leq \varepsilon^*(I') + \varepsilon^*(I'')$, keep I and forget all the preceding information about I' and I''; define $\varepsilon^*(I) = \varepsilon(I)$ and the partition of I is the trivial partition (consisting of only I).

(6.31)   If $\varepsilon(I) > \varepsilon^*(I') + \varepsilon^*(I'')$, set $\varepsilon^*(I) = \varepsilon^*(I') + \varepsilon^*(I'')$ and the partition of I is obtained by combining the partitions of I' and I'' that were used to calculate $\varepsilon^*(I')$ and $\varepsilon^*(I'')$.

Arriving at the "summit of the pyramid," that is to say, at L, we have, at the same time, found the minimal entropy and the optimal partition of L, which leads to the optimal basis.

6.8.   An example where this algorithm works.

Consider a signal of the form $g(t) + \frac{1}{\sqrt{h}} e^{i\omega t} g((t - t_0)/h)$, where $g(t) = e^{-t^2/2}$, $0 < h < 1$ and where ω is a real number, which can be arbitrarily large. We will be concerned with the limiting situation where h is very small.

If this signal $f(t)$ is analyzed using the Malvar wavelets associated with a regularly segmented grid $(a_j = ja)$, then the entropy of the decomposition is necessarily greater than $C \log 1/h$. Indeed, if the grid mesh is of order 1, the term $\frac{1}{\sqrt{h}} e^{i\omega t} g((t - t_0)/h)$ is very poorly represented, whereas if the mesh is of order h, the term $g(t)$ is very poorly represented.

We will show that the entropy of a decomposition of $f(t)$ can decrease to C (a constant) by using the adaptive segmentation of the last section.

We assume, to fix ideas, that $h = 2^{-q}$ and that the initial grid is $2^{-q}\mathbb{Z}$. The optimal partition in dyadic intervals is then formed from the sequence of nested dyadic intervals $J_q \subset J_{q-1} \subset \cdots \subset J_0$ containing t_0 and having lengths

$2^{-q}, 2 \cdot 2^{-q}, \ldots, 1$. To each J_n we associate the two contiguous intervals of the same length to the left and right of J_n. The extremities of the dyadic intervals thus defined constitute the optimal segmentation for $f(t)$.

It is then easy to show that the entropy of $f(t)$ in the Malvar basis corresponding to this segmentation does not exceed a certain constant C.

The adaptive segmentation has allowed us to "zoom in" on the singularity of $f(t)$, which is located at t_0. Thus, in this example, the optimal segmentation algorithm has provided an interesting analysis of the signal $f(t)$.

6.9. The discrete case.

We replace the real line $\mathbb{R}$ by the grid $h\mathbb{Z}$, where $h > 0$ is the sampling step. Thus the signal f is given by a sampling denoted $f(hk)$, $k = 0, \pm 1, \pm 2, \ldots$, but we will not discuss here the technique used to arrive at this sampling. We will systematically forget h in all that follows.

A partition of $\mathbb{Z}$ will be defined by the intervals $[a_j, a_{j+1}]$, where $a_j - \frac{1}{2}$ is an integer, in such a way that a_j does not belong to $\mathbb{Z}$. This point of view has also often been adapted in the context of the DCT.

We denote a number of points belonging to $[a_j, a_{j+1}] \cap \mathbb{Z}$ by $l_j = a_{j+1} - a_j$, and we let the numbers $\alpha_j > 0$ be small enough so that $\alpha_j + \alpha_{j+1} \le l_j$.

The windows $w_j(t)$ will be subject to exactly the same conditions as in the continuous case. This means that

$$(6.32) \qquad w_j(t) = 0 \quad \text{outside the interval} \quad [a_j - \alpha_j, a_{j+1} + \alpha_{j+1}]$$

$$(6.33) \qquad w_j(t) = 1 \quad \text{on the interval} \quad [a_j + \alpha_j, a_{j+1} - \alpha_{j+1}]$$

$$(6.34) \quad 0 \le w_j(t) \le 1 \quad \text{and} \quad w_{j-1}(a_j + \tau) = w_j(a_j - \tau) \quad \text{if } |\tau| \le \alpha_j$$

$$(6.35) \qquad w_j^2(a_j + \tau) + w_j^2(a_j - \tau) = 1 \quad \text{if} \quad |\tau| \le \alpha_j.$$

Then the double sequence

$$(6.36) \qquad \sqrt{\frac{2}{l_j}} w_j(t) \cos\left[\frac{\pi}{l_j}\left(k + \frac{1}{2}\right)(t - a_j)\right], \qquad 0 \le k \le l_j - 1, \qquad j \in \mathbb{Z},$$

is an orthonormal basis for $l^2(\mathbb{Z})$.

Furthermore, nothing prevents us from considering a finite interval of integers and replacing $l^2(\mathbb{Z})$ by $l^2\{1, \ldots, N\}$. Start with $a_0 = \frac{1}{2}$ and end with $a_{j_0+1} = N + \frac{1}{2}$. We require that $w_0(t)$ be equal to 1 on $[1/2, a_1 - \alpha_1]$, and there is no other constraint on this interval. Similarly, $w_{j_0}(t) = 1$ on $[a_{j_0+1} - \alpha_{j_0+1}, a_{j_0+1}]$ with no other constraint on the interval.

The Malvar wavelets thus exist in very different algorithmic settings, and it is this that makes them superior to other analytic techniques (Gabor wavelets, Grossmann–Morlet wavelets, etc.).

Bibliography

[1] R. COIFMAN AND Y. MEYER, *Remarques sur l'analyse de Fourier à fenêtre*, C. R. Acad. Sci. Paris Sér. I (1991), pp. 259–261.

[2] I. DAUBECHIES, *The wavelet transform, time-frequency localization and signal analysis*, IEEE Trans. Inform. Theory, 36 (1990), pp. 961–1005.

[3] S. E. ELNAHAS, KOU-HU TZOU, J. R. COX, R. L. HILL, AND R. G. JOST, *Progressive coding and transmission of digital diagnostic pictures*, IEEE Trans. Medical Imaging, MI-5, 2 (1986), pp. 73–83.

[4] H. S. MALVAR, *Lapped transforms for efficient transform/subband coding*, IEEE Trans. Acoust., Speech, Signal Process., 38 (1990), pp. 969–978.

[5] ———, *Fast algorithm for modulated lapped transform*, Electron. Lett., 27 (1991), pp. 775–776.

[6] ———, *Signal processing with lapped transforms*, Artech House, Norwood, MA, 1991.

[7] H. S. MALVAR AND D. H. STAELIN, *Reduction of blocking effects in image coding with a lapped orthogonal transform*, Proc. ICASSP 88, New York, April 1988, pp. 781–784.

[8] ———, *The LOT: Transform coding without blocking effects*, IEEE Trans. Acoust., Speech, Signal Process., 37 (1989), pp. 553–559.

CHAPTER 7

Time-Frequency Analysis and Wavelet Packets

7.1. Heuristic considerations.

A time-frequency analysis of a signal lets us express the signal as a linear combination of time-frequency atoms. These time-frequency atoms are essentially characterized by

(a) an arbitrary duration, $t_2 - t_1$; t_1 is the instant when the signal is first heard (if it is a speech signal, for example), and t_2 is the instant when it ceases to be heard, and

(b) an arbitrary frequency, which is the average frequency; this average frequency is the frequency of the emitted note in the case of a musical signal, while the frequency spectrum given by Fourier analysis takes into consideration the parasitic frequencies created by the note's attack and decay.

We also think of a time-frequency atom as occupying a symbolic region in the time-frequency plane. This symbolic region is a rectangle with area 2π (the Heisenberg uncertainty principle).

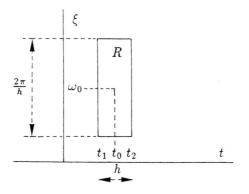

The most famous example of time-frequency atoms is that of the Gabor wavelets. For these we have $f_R(t) = e^{i\omega_0 t} g_h(t - t_0)$, where $t_0 = \frac{1}{2}(t_1 + t_2)$ is the center of the time-frequency atom and $g_h(t) = h^{-1/2} g(t/h)$, $g(t) = \pi^{-1/4} e^{-t^2/2}$.

To say that the time-frequency atom $f_R(t)$ occupies the symbolic region R of the time-frequency plane means that $f_R(t)$ is essentially supported by the interval

$[t_1, t_2]$ and that the Fourier transform $\hat{f}_R(\xi)$ of $f_R(t)$ is essentially supported by the interval $[\omega_0 - \pi/h, \ \omega_0 + \pi/h]$. It is well known that there does not exist a function with compact support whose Fourier transform also has compact support. This leads one to consider the following, less stringent conditions

$$(7.1) \qquad \int_{-\infty}^{\infty} (t - t_0)^2 |f_R(t)|^2 dt \leq C^2 h^2,$$

$$(7.2) \qquad \int_{-\infty}^{\infty} (\xi - \omega_o)^2 |\hat{f}_R(\xi)|^2 d\xi \leq 2\pi C^2 h^{-2}.$$

The time-frequency atom that optimizes this criterion (that is, for which the constant C is the smallest possible) is precisely the Gabor wavelet, and the Gabor wavelet owes it success to this optimal localization in the time-frequency plane.

We will see, on the other hand, that the Gabor wavelets have a very disagreeable property, which makes them unsuitable for the time-frequency signal analysis.

If the time-frequency atoms $f_R(t)$ were actually concentrated on rectangles R in the time-frequency plane, they would enjoy the following property: If R_1 and R_2 are disjoint rectangles in the time-frequency plane, then

$$(7.3) \qquad \int_{-\infty}^{\infty} f_{R_1}(t) \overline{f_{R_2}}(t) dt = 0.$$

We indicate the "proof" of this property. If R_1 and R_2 are disjoint, then either the horizontal sides of the rectangles are disjoint or the vertical sides are disjoint. In the first case, the supports of $f_{R_1}(t)$ and $f_{R_2}(t)$ (in t) are disjoint, and the integral (7.3) is zero. In the second case, the supports of the Fourier transforms $\hat{f}_{R_1}(\xi)$ and $\hat{f}_{R_2}(\xi)$ (in ξ) are disjoint, and the integral (7.3) is still zero, as we see by applying Parseval's identity. We know, if fact, that this cannot happen, and if $f_{R_1}(t)$ and $f_{R_2}(t)$ are Gabor wavelets, the integral (7.3) is never zero. But this integral is small if R_1 and R_2 are "remote," that is, if the rectangles mR_1 and mR_2 are disjoint. Here $m \geq 1$ is an integer, and mR is the rectangle that has the same center as R and whose sides are m times the length of the sides of R. If m is large, remoteness becomes a very strong condition.

Eric Séré [5] has shown that remoteness of the rectangles $R_0, R_1, R_2, \ldots$ does not imply that the corresponding Gabor wavelets $f_{R_0}(t), f_{R_1}(t), f_{R_2}(t), \ldots$ are well separated from each other. More precisely, for every m (no matter how large), there exist rectangles $R_0, R_1, \ldots$ in the time-frequency plane such that the rectangles mR_j are pairwise disjoint and coefficients $\alpha_0, \alpha_1, \ldots$, such that

$$(7.4) \qquad \sum_0^{\infty} |\alpha_j|^2 = 1,$$

and

$$(7.5) \qquad \int_{-\infty}^{\infty} \left| \sum_0^{\infty} \alpha_j f_{R_j}(t) \right|^2 dt = +\infty.$$

Thus remoteness of the rectangles in the time-frequency plane does not even imply that the corresponding Gabor wavelets are almost orthogonal, and consequently the apparent heuristic simplicity of the time-frequency plane is completely misleading.

This catastrophic phenomenon results from the arbitrariness of the $h > 0$ that are used in the definition of the time-frequency atoms. The rectangles $R_0, R_1, \ldots$ in Séré's result have arbitrarily large eccentricity. When $h = 1$ all is well, and the corresponding situation has been studied extensively. This is then a form of windowed Fourier analysis where the sliding window is a Gaussian [4].

Once we abandon the Gabor wavelets we have two options: Malvar wavelets and wavelet packets. We will briefly indicate the advantages and disadvantages of these two options.

If we use Malvar wavelets, then by their nature, the duration of the attack or of the decay is not related to the duration of the stationary part. We can, for example, have a Malvar wavelet for which the durations of the attack and of the decay are of order 1, while the stationary part lasts $T \gg 1$. If ω_0 is the frequency corresponding to this stationary part, then the Fourier transform of the Malvar wavelet will be, at best, of the form $\frac{\sin T(\xi - \omega_0)}{\sqrt{T}\,(\xi - \omega_0)}\hat{\varphi}(\xi - \omega_0)$, and it cannot satisfy (7.2) since h is of the order of magnitude of T. On the other hand, the Malvar wavelets are constructed to be exactly orthogonal.

The implication of our observations is that the orthogonality of the Malvar wavelets has been won at the price of their localization in the time-frequency plane, a localization that no longer guarantees the "minimal conditions" (7.1) and (7.2).

One last remark about the Malvar wavelets is obvious but significant: Although they are given by a very simple formula, they are not obtained by translation, change of scale, and modulation (or modulation by $e^{i\omega t}$) of a fixed function $g(t)$.

The option we propose in the following pages is that of wavelet packets. The advantages of this option are the following:

(a) Daubechies's orthogonal wavelets (Chapter 3) are a particular case of wavelet packets.

(b) Wavelet packets are organized naturally into collections, and each collection is an orthonormal basis for $L^2(\mathbb{R})$.

(c) Thus one can compare the advantages and disadvantages of the various possible decompositions of a given signal in these orthonormal bases and select the optimal collection of wavelet packets for representing the given signal.

(d) Wavelet packets are described by a very simple algorithm $2^{j/2} w_n(2^j x - k)$, where $j, k \in \mathbb{Z}$, $n = 0, 1, 2, \ldots$, and where the supports of the $w_n(x)$ are in the same fixed interval $[0, L]$.

The integer n plays the role of a frequency, and it can be compared to the integer $k = 0, 1, 2, \ldots$ that occurs in the definition of the Malvar wavelets.

The price to pay for these advantages is the same one that comes up when we use Malvar wavelets. Indeed, if, to facilitate intuition, we associate the rectangle R in the time-frequency plane defined by $k2^{-j} \leq t < (k+1)2^{-j}$ and $n2^j \leq \xi <$

$(n+1)2^j$ with the wavelet packet $2^{j/2}w_n(2^jt-k)$, then this choice does not meet the conditions (7.1) and (7.2). Furthermore, we cannot do better by assigning a frequency different from n to $w_n(x)$, for although $\|w_n\|_2 = 1$,

$$(7.6) \qquad \varlimsup_{n\to+\infty}\left\{\inf_{w\in\mathbb{R}}\int_{-\infty}^{\infty}(\xi-w)^2|\hat{w}_n(\xi)|^2d\xi\right\} = +\infty.$$

The frequency localization of wavelet packets is relatively poor, except for certain values of n, and hence the "lim sup" in (7.6).

7.2.   The definition of basic wavelet packets.

We begin by defining a special sequence of functions $w_n(x), n = 0, 1, 2, \dots$, supported by the interval $[0, 2N - 1]$, where $N \ge 1$ is fixed at the outset. If $N = 1$, these functions $w_n(x)$ constitute the Walsh system, which, we recall, is a well-known orthonormal basis for $L^2[0,1]$. If $N \ge 2$, the functions $w_n(x)$ are no longer supported by $[0,1]$; however, the double sequence

$$(7.7) \qquad w_n(x - k), \qquad n = 0, 1, 2, \dots, k \in \mathbb{Z}$$

will be an orthonormal basis for $L^2(\mathbb{R})$. This orthonormal basis will allow us to do an orthogonal windowed Fourier analysis. Thus, for the moment, this construction is similar to the Malvar wavelets. The difference occurs when the dilations enter (the changes of variable of the form $x \to 2^jx$).

We start with an integer $N \ge 1$ and consider two finite trigonometric sums

$$m_0(\xi) = \frac{1}{\sqrt{2}}\sum_0^{2N-1}h_ke^{-ik\xi}$$

$$(7.8)$$

$$m_1(\xi) = \frac{1}{\sqrt{2}}\sum_0^{2N-1}g_ke^{-ik\xi}$$

satisfying the following conditions:

$$(7.9) \qquad g_k = (-1)^{k+1}\overline{h}_{2N-1-k} \quad \text{or} \quad m_1(\xi) = e^{i(2N-1)\xi}\overline{m_0(\xi+\pi)},$$

$$(7.10) \qquad m_0(0) = 1, \quad m_0(\xi) \ne 0 \quad \text{on} \quad \left[-\frac{\pi}{3}, \frac{\pi}{3}\right],$$

and, finally,

$$(7.11) \qquad |m_0(\xi)|^2 + |m_0(\xi+\pi)|^2 = 1.$$

One choice is given by

$$(7.12) \qquad |m_0(\xi)|^2 = 1 = c_N\int_0^\xi(\sin t)^{2N-1}dt,$$

where

$$c_N \int_0^\pi (\sin t)^{2N-1} dt = 1,$$

but other choices are possible [2].

As a first example take $m_0 = \frac{1}{2}(e^{-i\xi} + 1)$ and $m_1(\xi) = \frac{1}{2}(e^{-i\xi} - 1)$. Condition (7.11) reduces to

$$\cos^2 \frac{\xi}{2} + \sin^2 \frac{\xi}{2} = 1.$$

A second choice is given by

$$\sqrt{2}\, h_0 = \frac{1}{4}(1 + \sqrt{3}), \qquad \sqrt{2}\, h_1 = \frac{1}{4}(3 + \sqrt{3}),$$

$$\sqrt{2}\, h_2 = \frac{1}{4}(3 - \sqrt{3}), \qquad \sqrt{2}\, h_3 = \frac{1}{4}(1 - \sqrt{3}),$$

The wavelet packets $w_n(x)$ are then defined by induction on $n = 0, 1, 2, \ldots,$ using the two identities

$$(7.13) \qquad w_{2n}(x) = \sqrt{2} \sum_0^{2N-1} h_k w_n(2x - k)$$

$$(7.14) \qquad w_{2n+1}(x) = \sqrt{2} \sum_0^{2N-1} g_k w_n(2x - k)$$

and by the condition $w_0(x) \in L^1(\mathbb{R})$, $\int_{-\infty}^\infty w_0(x) dx = 1$.

We explain the roles of these two identities. Identity (7.13), with $n = 0$, is

$$(7.15) \qquad w_0(x) = \sqrt{2} \sum_0^{2N-1} h_k w_0(2x - k),$$

and the function $\varphi(x) = w_0(x)$ appears as a fixed point of the operator $T : L^1(\mathbb{R}) \to L^1(\mathbb{R})$ defined by

$$(7.16) \qquad Tf(x) = \sqrt{2} \sum_0^{2N-1} h_k f(2x - k),$$

which becomes

$$(7.17) \qquad (Tf)\hat{\ }(\xi) = m_0(\xi/2)\hat{f}(\xi/2)$$

by taking the Fourier transform.

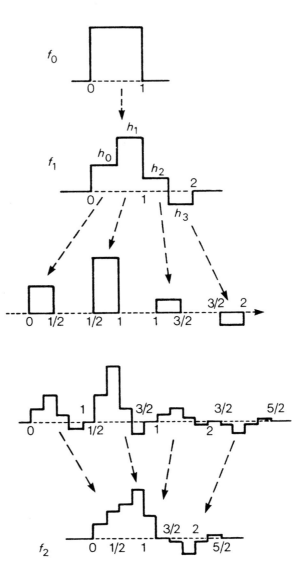

When f is normalized by $\int_{-\infty}^{\infty} f(x)dx = 1$, the fixed point is unique, and it is given by

(7.18) $\hat{\varphi}(\xi) = m_0(\xi/2)m_0(\xi/4)\ldots m_0(\xi/2^j)\ldots$

This diagram illustrates the iterative scheme for constructing φ using the characteristic function of $[0,1]$ for the initial value f_0. Here, $f_{j+1} = T(f_j)$, and we have drawn the first few functions f_0, f_1, and f_2. The coefficients h_0, h_1, h_2, and h_3 are approximately $\frac{1}{4\sqrt{2}}(1 + \sqrt{3}), \ldots, \frac{1}{4\sqrt{2}}(1 - \sqrt{3})$, as in the second example mentioned above. Then the sequence f_j converges uniformly to the fixed point $\varphi(x)$.

Once the function φ is constructed, we use (7.14) with $n = 0$, which gives us $\psi = w_1$. $\psi(x)$ is the "mother" wavelet in the construction of orthonormal wavelet bases, and $\varphi(x)$ is the "father" wavelet. Next, we use (7.13) and (7.14) with $n = 1$ and obtain $w_2(x)$ and $w_3(x)$. By repeating this process we generate, two at a time, all of the wavelet packets.

The support of $\varphi(x)$ is exactly the interval $[0, 2N - 1]$ (see [2]), and it is easy to show that the supports of $w_n(x), n \in \mathbb{N}$, are included in $[0, 2N - 1]$.

A very important result from this construction of wavelet packets is that the double sequence

$$(7.19) \qquad\qquad w_n(x - k), \qquad n = 0, 1, 2, \ldots, \qquad k \in \mathbb{Z},$$

is an orthonormal basis for $L^2(\mathbb{R})$.

To be more precise, the subsequence derived from (7.19) by taking $2^j \leq n < 2^{j+1}$ is an orthonormal basis for the orthogonal complement W_j of V_j in V_{j+1}. Recall that, in the language of multiresolution analysis, V_j is the closed subspace of $L^2(\mathbb{R})$ spanned by the orthonormal basis $2^{j/2}\varphi(2^j x - k)$, $k \in \mathbb{Z}$, and similarly, $2^{j/2}\psi(2^j x - k)$, $k \in \mathbb{Z}$, is an orthonormal basis for W_j. Thus, the construction of wavelet packets appears as a change of orthonormal basis inside each W_j.

An interesting observation concerns the case where the filter has length 1 and $h_0 = h_1 = \frac{1}{\sqrt{2}}$, $g_1 = -g_0 = \frac{1}{\sqrt{2}}$. Let us recall the definition of the Walsh system. We let $r(x)$ denote the periodic function with period 1 that equals 1 on the interval $[0, 1/2)$ and -1 on the interval $[1/2, 1)$. To define the Walsh system $W_n(x)$, $n \in \mathbb{N}$, let $\mathcal{X}(x)$ denote the characteristic function of the interval $[0, 1)$ and, for $n = \varepsilon_0 + 2\varepsilon_1 + \ldots + 2^j \varepsilon_j$, where $\varepsilon_j = 0$ or 1, write

$$W_n(x) = [r(x)]^{\varepsilon_0}[r(2x)]^{\varepsilon_1} \cdots [r(2^j x)]^{\varepsilon_j} \mathcal{X}(x).$$

It is helpful to observe that $r^0 = 1$ and that $r^1 = r$. Then we can immediately verify that

$$W_{2n}(x) = W_n(2x) + W_n(2x - 1)$$

and that

$$W_{2n+1}(x) = W_n(2x) - W_n(2x - 1).$$

This shows that, in the case of filters of length 1, the construction of wavelet packets leads to the Walsh system.

The Walsh system $W_n(x)$, $n \in \mathbb{N}$, is an orthonormal basis for $L^2[0, 1]$, and it follows immediately that the double sequence $W_n(x - k)$, $n \in \mathbb{N}$, $k \in \mathbb{Z}$, is an orthonormal basis for $L^2(\mathbb{R})$.

In the general case of basic wavelet packets (filters longer than 1), the supports of $w_n(x - k)$ and $w_{n'}(x - k')$ are not necessarily disjoint when $k \neq k'$ so that the orthogonality of the double sequence $w_n(x - k), n \in \mathbb{N}, k \in \mathbb{Z}$, is more subtle.

7.3.   General wavelet packets.

The basic wavelet packets are the functions $w_n(x), n = 0, 1, 2, \ldots$ (which are derived from a filter $\{h_k\}$), and the sequence $w_n(x - k), n \in \mathbb{N}, k \in \mathbb{Z}$, is an orthonormal basis for $L^2(\mathbb{R})$. This orthonormal basis is analogous to the Walsh system but, for filters longer than 1, it is more regular. That is, the frequency localization of the functions $w_n(x)$ is better than the frequency localization of the functions in the Walsh system. Nevertheless, this frequency localization does not yield an estimate of the type

$$(7.20) \qquad\qquad \inf_{\omega \in \mathbb{R}} \int_{-\infty}^{\infty} (\xi - \omega)^2 |\hat{w}_n(\xi)|^2 d\xi \leq C,$$

uniformly in n (see [2]).

The general wavelet packets are the functions

$$(7.21) \qquad\qquad 2^{j/2} w_n(2^j x - k), \qquad n \in \mathbb{N}, \qquad j, k \in \mathbb{Z}.$$

These are much too numerous to form an orthonormal basis. In fact, we can extract several different orthonormal bases from the collection (7.21). The choice $j = 0$, $n \in \mathbb{N}$, $j, k \in \mathbb{Z}$, leads to the orthonormal basis described in the previous section, while the choice $n = 1$, $j, k \in \mathbb{Z}$, leads to the orthonormal wavelet basis (Chapter 3).

We associate with each of the wavelet packets (7.21) the "frequency interval" $I, (j, n)$ defined by $2^j n \leq \xi < 2^j (n + 1)$. The following result describes certain sets of wavelet packets that make up orthonormal bases for $L^2(\mathbb{R})$.

THEOREM 7.1. *Let E be a set of pairs (j, n), $j \in \mathbb{Z}$, $n \in \mathbb{N}$, such that the corresponding frequency intervals $I(j, n)$ constitute a partition of $[0, \infty)$, up to a countable set. Then the subsequence*

$$(7.22) \qquad\qquad 2^{j/2} w_n(2^j x - k), \qquad (j, n) \in E, \qquad k \in \mathbb{Z},$$

is an orthonormal basis for $L^2(\mathbb{R})$.

Notice that choosing E is choosing a partition of the frequency axis. This partitioning is "active," whereas the corresponding sampling with respect to the variable x (or t) is passive and is dictated by Shannon's theorem.

Going back to Ville, wavelet packets lead to a signal analysis technique where the process is "first filter different frequency bands; then cut these bands into slices (in time) to study their energy variations."

Similarly, we refer to the methodology developed by Liénard: "The proposed analysis process contains the following steps: filtering with a zero-phase filterbank, and modeling the output signals into successive waveforms (channel-to-channel modeling)."

When we have at our disposal a "library" of orthonormal bases, each of which can be used to analyze a given signal of finite energy, we are necessarily faced with the problem of knowing which basis to choose. We settle this problem with the same approach that we used for the Malvar wavelets: The optimal choice is given by the entropy criterion that we have already used in the preceding chapter. This entropy criterion provides an adaptive filtering of the given signal.

7.4.   Splitting algorithms.

Let (α_k) and (β_k) be two sequences of coefficients, indexed by $k \in \mathbb{Z}$, and satisfying the following conditions: $\sum |\alpha_k|^2 < \infty$, $\sum |\beta_k|^2 < \infty$ and, by defining $m_0(\theta) = \sum \alpha_k e^{-ik\theta}$ and $m_1(\theta) = \sum \beta_k e^{-ik\theta}$, the matrix

$$U(\theta) = \begin{pmatrix} m_0(\theta) & m_1(\theta) \\ m_0(\theta + \pi) & m_1(\theta + \pi) \end{pmatrix} \quad \text{is unitary.}$$

Consider a Hilbert space H with an orthonormal basis $(e_k)_{k \in \mathbb{Z}}$, and define the sequence f_k, $k \in \mathbb{Z}$, of vectors in H by

$$(7.23) \qquad f_{2k} = \sqrt{2} \sum_{-\infty}^{\infty} \alpha_{2k-l} e_l, \qquad f_{2k+1} = \sqrt{2} \sum_{-\infty}^{\infty} \beta_{2k-l} e_l.$$

Then the sequence (f_k), indexed by $k \in \mathbb{Z}$, is also an orthonormal basis for the Hilbert space H.

Next, let H_0 be the closed subspace of H generated by the vectors f_{2k}, which we denote by $e_k^{(0)}$; similarly H_1 will be generated by $f_{2k+1} = e_k^{(1)}$, $k \in \mathbb{Z}$.

Nothing prevents us from repeating on $(H_0, e_k^{(0)})$ the operation we have done on (H, e_k) and from iterating these decompositions, while keeping the same coefficients (α_k) and (β_k) at each step.

An elementary example helps us to understand the nature of this splitting algorithm. The initial Hilbert space is $L^2[0, 2\pi]$ with the usual orthonormal basis $e_k = \frac{1}{\sqrt{2}} e^{ik\theta}$, $k \in \mathbb{Z}$. The (2π-periodic) functions $m_0(\theta)$ and $m_1(\theta)$ are (when restricted to $[0, 2\pi)$) the characteristic functions of $[0, \pi)$ and $[\pi, 2\pi)$. Then the vectors f_{2k}, $k \in \mathbb{Z}$, are $\frac{1}{\sqrt{\pi}} e^{i2k\theta} m_0(\theta)$ and constitute a Fourier basis for the interval $[0, \pi)$, while the vectors f_{2k+1}, $k \in \mathbb{Z}$, constitute a Fourier basis for the interval $[\pi, 2\pi)$. Finally, the subspace H_0 of H is composed of the functions supported on the interval $[0, \pi)$, while H_1 is composed of the functions supported on $[\pi, 2\pi)$.

Iterating the splitting algorithm leads to subspaces that are naturally denoted by $H_{(\varepsilon_1, \ldots, \varepsilon_j)}$, $\varepsilon_1 = 0$ or $1, \ldots$, $\varepsilon_j = 0$ or 1, or even by H_I, where I denotes the dyadic interval of length 2^{-j} and origin $\varepsilon_1/2 + \cdots + \varepsilon_j/2^j$. In fact, in the example we have just studied, H_I is exactly the subspace of $L^2[0, 2\pi]$ consisting of the functions that vanish outside the interval $2\pi I$.

This example has guided the intuition of scientists working in signal processing. Assuming that the signal is sampled on $\mathbb{Z}$, they have considered the situation where (α_k) and (β_k) are two finite sequences and where $m_0(\theta)$ resembles the transfer function of a low-pass filter while $m_1(\theta)$ resembles that of a high-pass filter. One requires, at least, that $m_0(0) = 1$ and that $m_0(\theta)$ does not vanish in the interval $[-\frac{\pi}{3}, \frac{\pi}{3}]$.

By analogy with the preceding example, these scientists were led to believe that the iterative scheme, which we have called the splitting algorithm, would provide a finer and finer frequency definition, as one wanders through the maze of "channels" in the following figure.

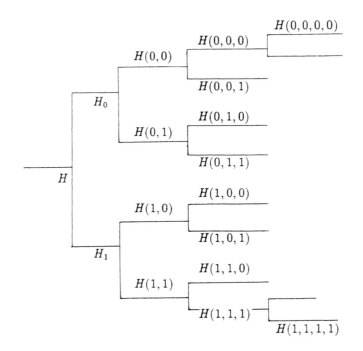

The initial Hilbert space H is the direct sum of various combinations of these subspaces. In particular, H is the direct sum of all the subspaces at the same "splitting level": at the first level there are 2 subspaces, at the next level there are 4, then 8, 16, and so on.

To give a better understanding of the construction of wavelet packets and the exact nature of the splitting algorithm, consider the case where the initial Hilbert space is the space V_j, $j \geq 1$, (in the language of multiresolution analysis) with the orthonormal basis $2^{j/2}\varphi(2^j x - k)$, $k \in \mathbb{Z}$. Next, suppose that the splitting algorithm has operated j times. Then we arrive exactly at the sequence of functions $w_n(x - k)$, $k \in \mathbb{Z}$, $0 \leq n < 2^j$. More precisely, the integer $n = \varepsilon_0 + 2\varepsilon_1 + \cdots + 2^{j-1}\varepsilon_{j-1}$ is the index of the "frequency channel" $H_{(\varepsilon_0, \varepsilon_1, \ldots, \varepsilon_{j-1})}$.

The frequency localization of wavelet packets does not conform to the intuition of the scientists who introduced these algorithms, and the only case where there is a precise relation between the integer n and a frequency in the sense of Fourier analysis is the case where $m_0(\theta)$ and $m_1(\theta)$ are the transfer functions of "ideal filters."

7.5. Conclusions.

It remains for us to indicate how to use wavelet packets. We begin by selecting, for use throughout the discussion, two sequences h_k, g_k, $0 \leq k \leq 2N - 1$, that satisfy the conditions for constructing wavelet packets. The choice of these sequences results from a compromise between the length $(2N - 1)$ of the filters and the quality of the frequency resolution. Once the filters are selected, we set in motion the algorithm for constructing the wavelet packets. From this processs we obtain a huge collection of orthonormal bases for $L^2(\mathbb{R})$.

It is then a question of determining, for a given signal, the optimal basis. And again, the optimal basis is the one (among all those in the wavelet packet) that gives the most compact decomposition of the signal.

We determine this optimal basis by using a "fine-to-coarse" type strategy and the method of merging. Thus we start from the finest frequency channels H_I, which are associated with the dyadic intervals I of length $|I| = 2^{-m}$. The integer m is taken to be as large as possible, consistent with the chosen precision. The algorithm proceeds by making the following decision: It combines the left and right halves, I' and I'', of a dyadic interval I whenever the orthonormal basis of H_I yields a more compact representation than that obtained by using the two orthonormal bases of $H_{I'}$ and $H_{I''}$.

The discrete version of wavelet packets can also be used and is immediately available. This is obtained by starting with the Hilbert space $H = l^2(\mathbb{Z})$ of signal sampled on $\mathbb{Z}$ and the canonical orthonormal basis $(\varepsilon_k)_{k\in\mathbb{Z}} : \varepsilon_k$ is 1 at k and 0 elsewhere. Here there is perfect resolution in position but no resolution in the frequency variable. Next, we systematically apply the splitting algorithm to improve the frequency definition until reaching the spaces H_I associated with the dyadic intervals I of length $|I| = 2^{-m}$. Finally, we apply the algorithm to choose the best basis (§6.7).

Wavelet packets offer a technique that is dual to the one given by the Malvar wavelets. In the case of wavelet packets, we effect an adaptive filtering, whereas the Malvar wavelets are associated with an adaptive segmentation.

Bibliography

[1] R. COIFMAN, *Adapted multiresolution analysis, computation, signal processing and operator theory*, ICM 90, Kyoto, Japan, Springer-Verlag.

[2] R. COIFMAN, Y. MEYER, AND V. WICKERHAUSER, *Size properties of wavelet packets*, in Wavelets and Their Applications, M. B. Ruskai, G. Beylkin, R. Coifman, I. Daubechies, S. Mallat, Y. Meyer, L. Raphael, eds., Jones & Bartlett, Boston, MA, 1992, pp. 453–470.

[3] R. COIFMANN, Y. MEYER, S. QUAKE, AND V. WICKERHAUSER, *Signal processing and compression with wavelet packets*, preprint, Yale University, New Haven, CT, 1990.

[4] I. DAUBECHIES, *The wavelet transform, time-frequency localization and signal analysis*, IEEE Trans. Inform. Theory, 36 (1990), pp. 961–1005.

[5] E. SÉRÉ, Thesis, University of Paris-Dauphine, CEREMADE, Paris, France, 1992.

CHAPTER 8

Computer Vision and Human Vision

We propose, in this chapter, to describe and comment on a small part of David Marr's work. We limit our discussion to Marr's analysis of the "low-level" processing of luminous information by the retinal cells. Marr hypothesized that the coding of this luminous information was done by using the zero-crossings of the wavelet transform. This hypothesis leads us to state the famous "Marr conjecture" and then to state its precise form as conjectured by Mallat. This precise form yields a remarkably effective algorithm. We will see, however, that the Mallat conjecture is incorrect, and this poses some fascinating new problems.

8.1. Marr's program.

Vision, A Computational Investigation into the Human Representation and Processing of Visual Information appeared in 1982, and it is somewhat analogous to Descartes's *Discours de la méthode.* Exactly as Descartes did, Marr takes us into his confidence and speaks to us as if we were one of his friends or colleagues from his laboratory. Marr confides to us his intellectual progress and tells us about his doubts, his hopes, and his enthusiasms. He gives a lively description of the theories he has struggled with and rejected, and he explains his own research program to the reader with a sense of jubilation.

We recall that the goal of Marvin Minsky's group at the Massachusetts Institute of Technology (MIT) artificial intelligence laboratory was to solve the problem of artificial vision for robots. The challenge was to construct a robot endowed with a perception of its environment that enabled it to perform specific tasks. It turned out that the first attempts to construct a robot capable of understanding its surroundings were completely unsuccessful.

These surprising setbacks showed that the problem of artificial vision was much more difficult that it seemed. The idea then occurred to imitate, within the limits imposed by the technology of robotics, certain solutions found in nature. Marr, who was an expert on the human visual system, was invited to join the MIT group and leave Cambridge, England, for Cambridge, Massachusetts. According to Marr, the disappointments of the robotics scientists were due to having skipped a step. They had tried to go directly from the statement of the problem to its solution without having at hand the basic scientific understanding that is necessary to construct effective algorithms.

Marr's first premise is that there exists a science of vision, that it must be developed, and that once there has been sufficient progress, the problems posed by vision in robots can be solved.

Marr's second premise is that the science of human vision is no different from the science of robotic (or computer) vision.

Marr's third premise is that it is as vain to imitate nature in the case of vision as it would have been to construct an airplane by imitating the form of birds and the structure of their feathers. On the other hand, he notes that the laws of aerodynamics explain the flight of birds and enable us to build airplanes.

Thus it is important, as much for human vision as for computer vision, to establish scientific foundations rather than blindly seeking solutions.

To develop this basic science, one must carefully define the scope of inquiry. In the case of human vision, one must clearly exclude everything that depends on training, culture, taste, etc., for instance, the ability to distinguish the canvas of a master from that of an imitator. One retains only the mechanical or involuntary aspects of vision, that is, those aspects that enable us to move around, to drive a car, and so on.

We thus limit the following discussion to low-level vision: This is the aspect of vision than enables us to recreate the three-dimensional organization of the physical world around us from the luminous excitations that stimulate the retina.

The notion that low-level vision functions according to universal scientific algorithms seemed, to some, an implausible idea, and it encountered two kinds of opposition. In the first place, neurophysiologists had discovered certain cells having specific visual functions. But Marr objected to this reductionist approach to the problems of vision, and he offered two criticisms on this subject:

(a) After several very stimulating discoveries, neurophysiologists had not made sufficient progress to enable them to explain the action of the human visual system based on a collection of *ad hoc* cells.

(b) It would be absurd to look for the cell that lets you immediately recognize your grandmother.

On another front, Marr objected to attempts by psychologists to relate the performance of the human visual system to a learning process: We recognize the familiar objects of our environment by dint of having seen and touched them simultaneously. Bela Julesz had made a fundamental discovery that eliminated this as a working hypothesis.

Julesz made a systematic study of the response of the human visual system when it was presented with completely artificial images—synthetic images having no significance—which were computer-generated, random-dot stereograms. If these synthetic images presented a certain "formal structure" that stimulated stereo-vision reflexes, the eye deduced, in several milliseconds and without the slightest hesitation, a three-dimensional organization of the image. This organization "in relief" is clearly only a mirage in which the mechanism of stero-vision finds itself trapped. This mechanism acts with the same speed, the same quality, and the same precision as if it were a matter of recognizing familiar objects. Thus familiarity with the objects that one sees plays no role in the primary mechanisms of vision. Marr set out to understand the algorithmic architecture of these mechanisms.

This venture can be compared to that of the 17th century physiologists who studied the human body by comparing it with a complex and subtle machine—an assembly of bones, joints, and nerves whose functioning could be explained, calculated, and predicted by the same laws that applied to winches and pulleys. A century and a half later, Claude Bernard made a similar connection between the organic functioning of the human body and progress in the nascent field of organic chemistry. The synthesis of urea (Wöhler, 1828) again reduced the gap between the chemistry of life and organic chemistry.

In their scientific approach, these researchers relied on solid, well-founded knowledge, which came either from mechanics or from chemistry. They then tried to effect a technology transfer and to apply results acquired in the study of matter to life science.

But what Marr set out to do was much more difficult because the relevant knowlege base, namely, an understanding of robotics, was too tenuous to serve as the nucleus for an explanation of the human visual system.

Marr asserts that the problems posed by human vision or by computer vision are of the same kind and that they are part of a coherent and rigorous theory, an articulate and logical doctrine.

It is advisable, at the outset, to set aside any consideration of whether the results will ultimately be implemented with copper wires or nerve cells and to limit the investigation to the four properties of human vision that we wish to imitate or reproduce in robots. These are

(a) the recognition of contours of objects, the contours that delimit objects and structure the environment into distinct objects,

(b) the sense of the third dimension from two-dimensional retinal images and the ability to arrive at a three-dimensional organization of the physical world,

(c) the extraction of relief from shadows, and

(d) the perception of movement in an animated scene.

The fundamental questions posed by Marr are the following:

• How is it scientifically possible to define the contours of objects from the variations of their light intensity?

• How is it possible to sense depth?

• How is movement sensed? How do we recognize that an object has moved by examining a succession of images?

Marr opened a very active area of contemporary scientific research by giving each of these problems a precise algorithmic formulation and by furnishing parts of the solution in the form of algorithms.

Marr's working hypothesis is that human vision and computer vision face the same problems. Thus the algorithmic solutions can and must be tested within the framework of robotics technology and artificial vision.

In case of success, it is necessary to investigate whether these algorithms are physiologically realistic. For example, Marr did not believe that neuronal circuits used iterative loops, which are an essential aspect of the existing algorithms.

This discussion raises the basic problem of knowing the nature of the *representation* (the terminology is due to Marr) on which the algorithms act. Marr

uses a simple comparison to help us understand the implications of a representation. If the problem at hand was adding integers, then the *representation* of
the integers could be given in the Roman system, in the decimal system, or in
the binary system. These three systems provide three *representations* of the integers. But the algorithms used for addition will be different in the three cases,
and they will vary greatly in difficulty. This shows that the choice of this or that
representation involves significant consequences.

Marr was truly fascinated by the essential and subtle role played by a representation. He wrote: "A representation, therefore, is not a foreign idea at all—we
use representations all the time. However, the notion that one can capture some
aspect of reality by making a description of it using a symbol and that to do so
can be useful seems to me a fascinating and powerful idea. . . "

8.2. The theory of zero-crossings.

Marr felt that image processing in the human visual system has a complex hierarchical structure, involving several layers of processing. The "low-level processing" furnishes a *representation* that is used by later stages of visual information
processing. Based on a very precise analysis of the functioning of the ganglion
cells, Marr was led to this hypothesis: The basic representation ("the raw primal
sketch") furnished by the retinal system is a succession of sketches at different
scales and these scales are in geometric progression. These sketches are made
with lines, and these lines are the famous zero-crossings that Marr uses in the
following argument:

> The first of the three stages described above concerns the detection of intensity
> changes. The two ideas underlying their detection are (1) that intensity changes
> occur at different scales in an image, and so their optimal detection requires the
> use of operators of different sizes; and (2) that a sudden intensity change will
> give rise to a peak or trough in the first derivative or, equivalently, to a zero-
> crossing in the second derivative. These ideas suggest that in order to detect
> intensity changes efficiently, one should search for a filter that has two salient
> characteristics. First and foremost, it should be a differential operator, taking
> either a first or second derivative of the image. Second, it should be capable of
> being tuned to act at any desired scale, so that large filters can be used to detect
> blurry shadow edges, and small ones to detect sharply focused fine details in the
> image.
>   Marr and Hildreth (1980) argued that the most satisfactory operator ful-
> filling those conditions is the filter ΔG, where Δ is the Laplacian operator
> $\partial^2/\partial x^2 + \partial^2/\partial y^2$ and G stands for the two-dimensional Gaussian distribution
> $G(x,y) = e^{-(x^2+y^2)/2\sigma^2}$, which has standard deviation of $\sigma\sqrt{\pi}$. ΔG is a circu-
> larly symmetric Mexican-hat-shaped operator whose distribution in two dimen-
> sions may be expressed in terms of the radial distance r from the origin by the
> formula

$$\Delta G(r) = -\frac{2}{\sigma^2}\left(1 - \frac{r^2}{2\sigma^2}\right)e^{-r^2/2\sigma^2} \dots$$

We observe that $\Delta G(r) = (1/\sigma^2)\psi(\frac{x}{\sigma}, \frac{y}{\sigma})$, where $\psi(x,y)$ is the wavelet that
everyone today calls the Marr wavelet. If a black and white image is defined by

the gray levels $f(x,y)$, the zero-crossings of Marr's theory are the lines of the equation $(f * \psi_\sigma)(x,y) = 0$. Since the function $\psi(x,y)$ is even, the convolution product $(f * \psi_\sigma)(x,y)$ is (up to a proportionality factor) the wavelet coefficient of f, analyzed with the wavelet ψ. Thus the zero-crossings are defined by the vanishing of the wavelet coefficients.

It remains to specify the values of σ. These values, in geometric progression, were discovered by Campbell, Robson, Wilson, Giese, and Bergen, based on neurophysiological experiments. These experiments led to the values $\sigma_j = (1.75)^j \sigma_0$.

Marr's conjecture is that the original image $f(x,y)$ is completely determined by the sequence of lines where the functions $(f * \psi_{\sigma_j})(x,y)$ are zero. Interest in this representation of an image stems from its invariance under translations, rotations, and dilations (by powers of 1.75).

We quote Marr: "Zero-crossings provide a natural way of moving from an analogue or continuous representation like the two-dimensional image intensity values $I(x,y)$ to a discrete, symbolic representation. A fascinating thing about this representation is that it probably incurs no loss of information. The arguments supporting this are not yet secure..."

In the following pages, we propose to study Marr's conjecture. We will show first of all that it is incorrect for periodic images covering an unbounded idea. This does not exclude the possibility that the conjecture is true for bounded images, that is, images having finite extent.

We will then examine Mallat's conjecture, which is a version of Marr's conjecture. Mallat's conjecture provides an explicit algorithm for reconstructing the image. This algorithm works very well in spite of the fact that Mallat's conjecture is false. The counterexample that we construct is, in a certain sense, more realistic than the one we present in the case of Marr's conjecture. This counterexample raises an exciting problem: Why does Mallat's algorithm work so well?

8.3. A counterexample to Marr's conjecture.

We begin with a counterexample in one dimension. It will then be easy to transform it into a two-dimensional counterexample. This counterexample has the property of being periodic in x (or in x and y in the two-dimensional case). We do not know how to construct other counterexamples.

Consider all the functions $f(x)$ of the real variable x, having real values, and given by the series

$$(8.1) \qquad f(x) = \sin x + \sum_{2}^{\infty} \alpha_k \sin kx,$$

where

$$(8.2) \qquad \sum_{2}^{\infty} k^3 |\alpha_k| < 1.$$

We are going to show that all choices of the coefficients α_k lead to the same zero-crossings. For example $\sin x$ and $\sin x + \frac{1}{9} \sin 2x$ have the same zero-crossings. We verify this assertion by systematically applying the following simple observation: If $u(x)$ and $v(x)$ are two continuous functions of x, and if, for some constant $r \in [0, 1)$, $|v(x)| \leq r|u(x)|$ for all x, then $u(x) + v(x) = 0$ is equivalent to $u(x) = 0$.

Returning to (8.1), we define $g_\delta(x) = \frac{1}{\delta\sqrt{2\pi}}e^{-x^2/2\delta^2}$. Then

$$f * g_\delta(x) = e^{-\delta^2/2}\sin x + \sum_2^\infty \alpha_k e^{-k^2\delta^2/2}\sin kx.$$

It follows from this that

$$-\frac{d^2}{dx^2}(f * g_\delta)(x) = e^{-\delta^2/2}\sin x + \sum_2^\infty k^2\alpha_k e^{-k^2\delta^2/2}\sin kx$$

$$= u(x) + v(x).$$

Observe that $|\sin kx| \leq k|\sin x|$, which implies that $|v(x)| \leq r|u(x)|$, where $r = \sum_2^\infty k^3|\alpha_k| < 1$. Thus the zero-crossings of all the functions $f(x)$ are $x = m\pi$, $m \in \mathbb{Z}$.

If we wish to have $0 \leq f(x) \leq 1$, it is sufficient to add a suitable constant to $f(x)$ (defined by (8.1)) and then to renormalize the result by multiplication with a suitable positive constant. These two operations do not change the positions of the zero-crossings.

A nontrivial two-dimensional counterexample is

$$f(x, y) = \sin x \sin y + \sum_2^\infty \alpha_k \sin kx \sin ky,$$

where

$$2\sum_2^\infty k^4|\alpha_k| < 1.$$

8.4. Mallat's conjecture.

The existence of these counterexamples and several remarks Marr made in his book led Stephane Mallat to a more precise version of Marr's conjecture. He gave it a formulation that was compatible with the progress made in the early 1980s on pyramid algorithms for numerical image processing.

Mallat observed that numerical image processing using certain kinds of pyramid algorithms (quadrature mirror filters) and Marr's approach represented two particular examples of wavelet analysis of an image.

In fact, one has $\Delta(f * g_\delta) = \delta^{-2}f * \psi_\delta$, where

$$\psi(x, y) = -\frac{1}{\pi}\left(1 - \frac{x^2 + y^2}{2}\right)\exp\left(-\frac{x^2 + y^2}{2}\right)$$

is Marr's wavelet. With this in mind, Mallat took up a promising approach: to give Marr's conjecture a precise numerical and algorithmic content by taking advantage of the progress that had been made using quadrature mirror filters.

We start with the one-dimensional case. Mallat replaced the Gaussian $\frac{1}{\sqrt{2\pi}}e^{-x^2/2}$ with the basic cubic spline $\theta(x)$, whose support is the interval $[-2,2]$. Recall that $\theta = T * T$, where $T(x)$ is the triangle function that is equal to $1 - |x|$ if $|x| \leq 1$ and 0 if $|x| > 1$.

Let $f(x)$ be the function we wish to analyze by the method of zero-crossings and write $\theta_\delta(x) = \delta^{-1}\theta(\delta^{-1}x)$. Then the zero-crossings are the values of x where the second derivative $(d^2/dx^2)(f * \theta_\delta)$ is zero and changes sign.

To use the pyramid algorithms, Mallat assumes that $\delta = 2^{-j}$, $j \in \mathbb{Z}$. Mallat then proposes to code the signal $f(x)$ with the double sequence $(x_{q,j}, y_{q,j})$, where

(a) $x = x_{q,j}$ is (for $\delta = 2^{-j}$) a zero, with change of sign, of $(d^2/dx^2)(f * \theta_\delta)(x)$, and

(b) $y_{q,j} = \frac{d}{dx}(f * \theta_\delta)(x_{q,j})$.

In other words, Mallat considers the values of x (denote by $x_{q,j}$), where $\frac{d}{dx}(f * \theta_\delta)$ has an extremum, and he keeps the values of these local extrema in memory.

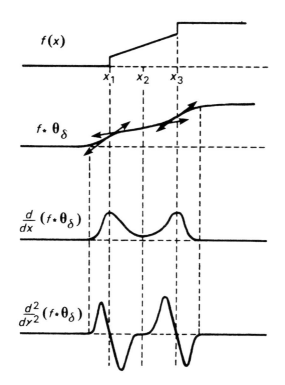

Certain of these local extrema correspond to points where the signal $f(x)$ changes rapidly; this is the case for the points x_1 and x_3 in the figure. Other extrema are related to points where the function changes very little. Mallat had the idea to consider only the first of these and thus to retain only the local maxima of $|\frac{d}{dx}(f * \theta_\delta)|$. This will not change the critical analysis that follows.

Coding f with the double sequence $(x_{q,j}, y_{q,j})$ meets two objectives: It is invariant under translation, and it corresponds to a precise form of Marr's conjecture. One reads on page 68 of his book: "On the other hand, we do have extra information, namely, the values of the slopes of the curves as they cross zero, since this corresponds roughly to the contrast of the underlying edge in the image. An analytic approach to the problem seems to be difficult, but in an empirical investigation, Nishihara (1981) found encouraging evidence supporting the view that a two-dimensional filtered image can be reconstructed from its zero-crossings and their slopes."

We are going to show that this conjecture is incorrect. However, this assertion must be tempered because our counterexample depends on a specific choice for the function $\theta(x)$. If $\theta(x)$ is the cubic spline, then we have a counterexample. If, on the other hand, the cubic spline $\theta(x)$ is replaced with the function that is equal to $1 + \cos x$ if $|x| \leq \pi$ and to 0 if $|x| > \pi$ (which is the Tukey window), then, for all signals $f(x)$ with compact support, reconstruction is theoretically possible but unstable. In this case, it comes down to determining a function with compact support from the knowledge of its Fourier transform in the neighborhood of zero, which is an unstable problem.

To describe the counterexample, we make a change of scale so that the values of δ are $2\pi 2^{-j}$ rather than 2^{-j}, $j \in \mathbb{Z}$.

Then we define $f_0(x) = 1 + \cos x$ if $-\pi \leq x \leq \pi$ and $f_0(x) = 0$ if $|x| \geq x$. We next consider the function $f_1(x)$ defined by

$$f_1(x) = f_0(x) + \sum_0^\infty \alpha_k (1 + \cos(2k+1)x) \quad \text{if} \quad -\pi \leq x \leq \pi$$

and

$$f_1(x) = 0 \quad \text{if} \quad |x| \geq x.$$

We choose the coefficients α_k, $k \geq 0$, so that

(8.3) $$|\alpha_k| k^m \to 0 \quad \text{as} \quad k \to +\infty \quad \text{for all} \quad m \geq 1,$$

(8.4) $$\sum_0^\infty (2k+1)^3 |\alpha_k| < \frac{1}{8}.$$

(8.5) $$\sum_0^\infty (-1)^k (2k+1)^{-3} \alpha_k \cos\left[(2k+1)\tau\right] = 0 \quad \text{for all} \quad \tau = 2\pi 2^{-j}, \quad j \in \mathbb{Z},$$

and finally a finite set of conditions of the form

(8.6) $$\sum_0^\infty \alpha_k \beta_{k,l} = 0, \quad 1 \leq l \leq N, \quad \sup_k |\beta_{k,l}| \leq C.$$

Then if $0 \leq \delta \leq \frac{\pi}{16}$, $(d^2/dx^2)(f_1 * \theta_\delta)$ and $(d^2/dx^2)(f_0 * \theta_\delta)$ have the same zeros, namely, at $x = \pm\frac{\pi}{2}$ or at $|x| \geq \pi + 2\delta$. The fact that the derivatives of $f_1 * \theta_\delta$ and $f_0 * \theta_\delta$ are the same at these zero-crossings results from (8.5).

Determining the zero-crossings is also easy if $\delta \geq 8\pi$; they are given by $x = \pm\frac{2\delta}{3}$ or at $|x| \geq \pi + 2\delta$. Finally, in the intermediate cases where $-1 \leq j \leq 4$, condition (8.6) forces the zero-crossings of $(d^2/dx^2)(f_1 * \theta_\delta)$ and $(d^2/dx^2)(f_0 * \theta_\delta)$ to be at the same places and the values of the first derivatives to be equal at these points.

The counterexample that we have just given looks very much like the one given in connection with Marr's conjecture. Notice that the modifications added to $f_0(x)$ involve "high frequencies" and that they are small.

If, however, the function $f(x)$ that we wish to analyze by Mallat's algorithm is a step function (with an arbitrarily large number of discontinuities), then Mallat's conjecture is correct. In fact, thanks to the symmetry of the function $\theta(x)$, the zero-crossings occur (for sufficiently small $\delta > 0$) at the points of discontinuity, while the values of the first derivatives of the smoothed signal furnish the jumps in the signal at these discontinuities. In this case, we have perfect reconstruction of the signal.

All this explains, without doubt, why Mallat's algorithm works in practice with such excellent precision, no matter what signals are treated. The signals in question have more in common with step functions than with the subtle functions described in the counterexamples.

8.5. The two-dimensional version of Mallat's algorithm.

We start with a two-dimensional image $f(x, y)$. From this we create the increasingly blurred versions at scales $\delta = 2^{-j}$, $j \in \mathbb{Z}$, by taking the various convolution products $f * \theta_\delta$, where, in two dimensions, $\theta_\delta(x, y) = \theta_\delta(x)\theta_\delta(y) = \delta^{-2}\theta(x/\delta)\theta(y/\delta)$. The function $\theta(x)$ is the basic cubic spline used in one dimension.

Next we consider the local maxima of the modulus of the gradient of $f * \theta_\delta$. We keep in memory the positions of these local maxima as well as the gradients at these points. The conjecture is that this data, computed for $\delta = 2^{-j}$, characterizes the image whose gray levels are given by $f(x, y)$.

We will show that this conjecture is incorrect in this general form. This does not exclude the possibility of its being true if (1) more restrictive assumptions are made about the function $f(x, y)$ or (2) the definition of the smoothing operator is changed.

The counterexample in two dimensions will be $f(x, y) = f_1(x) + f_1(y)$, where $f_1(x)$ is the counterexample in one dimension. One should observe here that

$$f_1(x) = f_0(x) + \sum_0^\infty \alpha_k(1 + \cos(2k + 1)x) \quad \text{for} \quad -\pi \leq x \leq \pi$$

and

$$\sum_0^\infty (2k + 1)^3 |\alpha_k| < \frac{1}{8}$$

imply that $\frac{7}{8} f_0(x) \leq f_1(x) \leq \frac{9}{8} f_0(x)$ since

$$1 + \cos(2k + 1)x \leq (2k + 1)^2(1 + \cos x).$$

The support of the function $f(x, y)$ is thus the cross $-\pi \leq x \leq \pi$ or $-\pi \leq y \leq \pi$ since

$$\frac{7}{8}(f_0(x) + f_0(y)) \leq f(x, y) \leq \frac{9}{8}(f_0(x) + f_0(y)).$$

Finally,

$$f(x, y) * \theta_\delta(x)\theta_\delta(y) = f_1(x) * \theta_\delta(x) + f_1(y) * \theta_\delta(y)$$

and the gradient of this function is the vector

$$\left(\frac{d}{dx}(f_1 * \theta_\delta)(x), \frac{d}{dy}(f_1 * \theta_\delta)(y)\right).$$

Its length is $(|\frac{d}{dx}(f_1 * \theta_\delta)|^2 + |\frac{d}{dy}(f_1 * \theta_\delta)|^2)^{1/2}$. It has a maximum if and only if $|\frac{d}{dx}(f_1 * \theta_\delta)|$ and $|\frac{d}{dy}(f_1 * \theta_\delta)|$ are at a maximum. But the set of functions $f_1(x)$ has been constructed so that the positions of the maxima of $|\frac{d}{dx}(f_1 * \theta_\delta)|$ are independent of the choice of f_1 and the same is true for the values of $\frac{d}{dx}(f_1 * \theta_\delta)$ at these points when $\delta = 2\pi 2^{-j}$, $j \in \mathbb{Z}$. Thus we have a counterexample in dimension two. Finding a counterexample whose support is a square is an important unsolved problem.

8.6. Conclusions.

All of this shows that Marr's conjecture is doubtful. However, the counterexample that we have constructed is deficient because it does not have finite support.

Regarding Mallat's conjecture, one must distinguish between the problem of unique representation and that of stable reconstruction. In our opinion, the reconstruction is never stable (unless the class of images to which the algorithm is applied is seriously limited). But it is, in certain cases, a representation that defines the image uniquely.

Bibliography

[1] S. G. MALLAT, *Multifrequency channel decompositions of images and wavelet models*, IEEE Trans. on Acoustics, Speech Signal Process., 37 (1989), pp. 2091–2110.

[2] S. G. MALLAT AND W.-L. HWANG, *Singularity detection and processing with wavelets*, Robotics Report No. 245, Courant Institute of Mathematical Sciences, New York, NY, March 1990.

[3] S. G. MALLAT AND S. ZHONG, *Complete signal representation with multiscale edges*, Robotics Report No. 219, Courant Institute of Mathematical Sciences, New York, NY, December 1989.

CHAPTER 9

Wavelets and Fractals

9.1. Introduction.

This title, in fact, pertains to the last three chapters of the book. In the next chapter, wavelets will be used to analyze the multifractal structure of fully developed turbulence. Similarly, in Chapter 11, the astronomer Albert Bijaoui will propose wavelets as a tool capable of elucidating the multifractal structure of the hierarchical organization of the galaxies.

In this chapter, we show that Fourier analysis does not easily reveal the multifractal structure of a signal and that wavelet analysis provides immediate access to information that is obscured by time-frequency methods. The two examples we present have played a significant role in the history of mathematics and, for this reason, this chapter has been written with extra care, by presenting all the details of the arguments and demonstrations. We invite the reader to accept our apologies, if he or she is not a mathematician, and to proceed directly to Chapter 10. We hope that readers who are mathematicians will enjoy reading this chapter as much as we have enjoyed writing it.

9.2. The Weierstrass function.

We intend to show that the function $\sigma(t) = \sum_0^\infty 2^{-j} \cos(2^j t)$ is nowhere differentiable and that the same is true for the function $\sum_0^\infty 2^{-j} \sin(2^j t)$. These proofs will use wavelet analysis, which, in this example, appears in general outline as a form of Littlewood–Paley analysis. The method we follow is due to Géza Freud.

Let $\psi(t)$ denote a function belonging to the holomorphic Hardy space $H^2(\mathbb{R})$. Assume, in addition, that $|\psi(t)| \leq C/|t|^3$ if $|t| \geq 1$ and that the Fourier transform $\hat{\psi}(\xi)$ of $\psi(t)$ satisfies the following conditions:

(9.1) $\quad \hat{\psi}(\xi) = 0 \quad$ if $\xi \leq 1/2$ (and in particular on $(-\infty, 0]$),

(9.2) $\quad \hat{\psi}(\xi) = 0 \quad$ if $\quad \xi \geq 2$,

(9.3) $\quad \hat{\psi}(1) = 1$.

We write $\psi_j(t) = 2^j \psi(2^j t)$ and denote the convolution operators $f \rightarrow f * \psi_j$ by Δ_j, $j = 0, 1, 2, \ldots$ These operators Δ_j, $j = 0, 1, 2, \ldots$ constitute a bank of filters that are arrayed on octave intervals. Thus the analysis of a real function f using the sequence Δ_j, $j = 0, 1, 2, \ldots$ resembles a Littlewood–Paley analysis that would be carried out on the analytic signal whose real part is f.

Freud's method is based on the following lemma.

LEMMA 9.1. *Let $f(t)$ be a bounded, continuous function of the real variable t. Assume that f is differentiable at t_0. Then*

$$\Delta_j f(t_0) = 2^{-j} \varepsilon_j \quad \text{where} \quad \varepsilon_j \to 0 \quad \text{as} \quad j \to +\infty.$$

By definition, $\Delta_j f(t_0) = 2^j \int f(t_0 - t)\psi(2^j t)dt$. We write $f(t_0 + t) = f(t_0) + tf'(t_0) + t\varepsilon(t)$, where $\varepsilon(t) \to 0$ with t and $|\varepsilon(t)| \le C$ if $|t| \ge 1$. This gives three terms for $\Delta_j f(t_0)$. The first two are zero because $\int \psi(t)dt = \int t\psi(t)dt = 0$. The third term is $2^j \int t\varepsilon(t)\psi(2^j t)dt = 2^{-j} \int \varepsilon(2^{-j}t)t\psi(t)dt$. But we have $|\varepsilon(2^{-j}t)| \le C$, $\lim_{j \to +\infty} \varepsilon(2^{-j}t) = 0$ (simple convergence), and $\int |t| \, |\psi(t)|dt < \infty$. From this it follows that $\varepsilon_j = \int \varepsilon(2^{-j}t)t\psi(t)dt \to 0$ as $j \to +\infty$.

Let's return to the two functions $\sigma(t) = \sum_0^\infty 2^{-j} \cos(2^j t)$ and $\tilde{\sigma}(t) = \sum_0^\infty 2^{-j} \sin(2^j t)$. Then by direct computation, $(\Delta_j \sigma)(t) = 2^{-j-1}e^{i2^j t}$ and $(\Delta_j \tilde{\sigma})(t) = -i2^{-j-1}e^{i2^j t}$. Lemma 9.1 applies, and the conclusion is that σ and $\tilde{\sigma}$ are nowhere differentiable.

We make the following observation about the choice of the analyzing wavelet $\psi(t)$. If we had chosen an even function ψ, decreasing as $1/|t|^3$ at infinity and satisfying $\hat{\psi}(\xi) = 0$ if $|\xi| \le 1/2$ or $|\xi| \ge 2$, this would have led to $(\Delta_j \tilde{\sigma})(t) = 2^{-j} \sin(2^j t)$, and we could not have concluded that $\hat{\sigma}$ is nondifferentiable when $t = p2^{-q}\pi$. From this example we see the merit of choosing an analyzing wavelet that is analytic. The information contained in $\Delta_j f(t)$ is more specific.

However, the choice of the analyzing wavelet loses all importance if we rephrase Lemma 9.1 in the following, more precise form.

LEMMA 9.2. *With the same hypotheses as Lemma 9.1, there exists a function $\eta(x)$, defined for $x \ge 0$, increasing, zero at $x = 0$, and continuous at 0, such that*

$$(9.4) \qquad |\Delta_j f(t_1)| \le |t_1 - t_0|\eta(|t_1 - t_0|) + 2^{-j}\eta(2^{-j})$$

for all $j \ge 0$ and all real t_1.

The proof is similar to the proof of Lemma 9.1.

We return to the issue of choosing the analyzing wavelet. By making the "bad choice" of a real, even wavelet, we ended up with $(\Delta_j \tilde{\sigma})(t) = 2^{-j} \sin(2^j t)$, and we were not able, using Lemma 9.1, to reach the desired conclusion. The result follows from Lemma 9.2 however. For example, for $t_0 = 0$, take $t_1 = \frac{\pi}{2}2^{-j}$ so that $\sin 2^j t_1 = 1$.

For experimental applications, users prefer a formulation such as Lemma 9.1 because there are considerably fewer values to check (one needs only to vary j), whereas the use of Lemma 9.2 requires one to consider (in addition to j) all values of t_1. This leads to a much more cumbersome algorithm.

The statement of Lemma 9.2 comes close to being a necessary and sufficient condition for differentiability at t_0. In fact, differentiability at t_0 cannot be characterized in this way, and the sufficient condition that we will present in §9.3 is not exactly the converse of Lemma 9.2

9.3. The determination of regular points in a fractal background.

We propose to determine the points t_0 where a function, which is otherwise very irregular, is differentiable. This is done in terms of a condition on the modulus of the wavelet coefficients. This condition is a sort of converse to Lemma 9.2.

We first state the result in terms of Daubechies's orthogonal wavelets with C^2 regularity. Next, we will indicate a "corollary of the proof," which will serve to analyze the differentiability of Riemann's function.

Let $\psi(t)$ be a function in class C^2, whose support is the interval $[0, L]$, and such that $2^{j/2}\psi(2^j t - k) = \psi_{j,k}(t)$, $j, k \in \mathbb{Z}$, is an orthonormal basis for $L^2(\mathbb{R})$.

The following lemma is the analogue of Lemma 9.2 in the language of wavelets.

LEMMA 9.3. *If $f(t)$ is a bounded, continuous function and if $f(t)$ is differentiable at t_0, then its wavelet coefficients $\alpha(j, k) = \int f(t)\psi_{j,k}(t)dt$ satisfy*

$$(9.5) \qquad |\alpha(j,k)| \leq 2^{-j/2}[2^{-j}\eta(2^{-j}) + |k2^{-j} - t_0|\eta(|k2^{-j} - t_0|)],$$

where $\eta(x)$ is defined for $x \geq 0$, is increasing, and is continuous at 0, and $\eta(0) = 0$.

The following theorem is a partial converse of Lemma 9.3. On one hand, we must make an assumption about the global regularity of the function $f(t)$, and for this we suppose that $f(t)$ belongs to the Hölder space C^α $(0 < \alpha < 1)$. On the other hand, we must strengthen the conditions on $\eta(t)$.

Recall that $f \in C^\alpha$ means that there exists a constant C such that for all t and all $h > 0$

$$(9.6) \qquad |f(t+h) - f(t)| \leq Ch^\alpha.$$

In terms of wavelet coefficients, this translates into the condition

$$(9.7) \qquad |\alpha(j,k)| \leq C2^{-j/2}2^{-j\alpha},$$

where here and elsewhere, the constant C may vary from expression to expression.

Note that only the values $h \leq 1$ are important in (9.6) since $f(t)$ is assumed to be bounded; if (9.6) holds for these values, then (9.7) is satisfied for all $j \geq 0$.

THEOREM 9.1. *Let f be a function of the real variable t that is in the class C^α, and let t_0 be a fixed real number. Assume that the wavelet coefficients $\alpha(j, k)$ of f satisfy, for $j \geq 0$ and all $k \in \mathbb{Z}$, the condition*

$$(9.8) \qquad |\alpha(j,k)| \leq 2^{-j/2}[2^{-j}\eta(2^{-j}) + |k2^{-j} - t_0|\eta(|k2^{-j} - t_0|)],$$

where $\eta(x)$ is defined for $x \geq 0$, is increasing, continuous at 0, and satisfies the Dini condition

$$(9.9) \qquad \int_0^1 \eta(x)\frac{dx}{x} < \infty.$$

Then f is differentiable at t_0, and $f'(t_0)$ can be computed by differentiating the wavelet expansion of f term by term.

Theorem 9.1 is not exactly the converse of Lemma 9.3 because we use two additional conditions: a global regularity hypothesis on f and the Dini condition (9.9).

The proof of Theorem 9.1 is so nice that we cannot resist the temptation to present it.

In our notation, both the constant C and the function η may vary from place to place. To be specific, when functions of the form $\eta(Kx) = \eta_K(x)$, $K \geq 0$, appear in estimates, we replace these (which are finite in number) by the largest of the η_K and rename it η. Note that all the η_K satisfy the same conditions as η.

By neglecting the contribution of the "low frequencies," which is a very regular function, we can write $f(t) = \sum_{j \geq 0} \sum_k \alpha(j,k) \psi_{j,k}(t)$. We then have

$$f(t_0 + h) - f(t_0) - h \sum_{j \geq 0} \sum_k \alpha(j,k) \psi'_{j,k}(t_0) = S_1(h) + S_2(h) + S_3(h),$$

where these three sums are taken over three intervals of j values.

Define the integer j_0 by $2^{-j_0} \leq \frac{|h|}{L} < 2 \cdot 2^{-j_0}$ and then define $j_1 \geq j_0$ by $2^{-j_1} \leq |h|^{2/\alpha} < 2 \cdot 2^{-j_1}$. Write

$$(9.10) \qquad S_1(h) = \sum_{0 \leq j \leq j_0} \sum_k \alpha(j,k) [\psi_{j,k}(t_0 + h) - \psi_{j,k}(t_0) - h\psi'_{j,k}(t_0)],$$

$S_2(h)$ is the analogous sum taken over $j_0 < j \leq j_1$, and $S_3(h)$ involves the values $j > j_1$.

We use two facts to estimate $|S_1(h)|$. On one hand, the functions $\psi_{j,k}(t)$ are sufficiently regular so that

$$(9.11) \qquad |\psi_{j,k}(t_0 + h) - \psi_{j,k}(t_0) - h\psi'_{j,k}(t_0)| \leq Ch^2 2^{5j/2},$$

and, on the other hand, the support of $\psi_{j,k}$ is the interval $[k2^{-j}, (k+L)2^{-j}]$. Once t_0 and h are fixed, the number of values of k for which $\psi_{j,k}(t_0)$ and $\psi_{j,k}(t_0 + h)$ are not zero is no greater than $3L$. For these values of k, (9.8) reduces to $|\alpha(j,k)| \leq C2^{-3j/2}\eta(2^{-j})$ since $|k2^{-j} - t_0| \leq C2^{-j}$. Finally,

$$(9.12) \qquad |S_1(h)| \leq Ch^2 \sum_{0 \leq j \leq j_0} 2^j \eta(2^{-j}) = o(h),$$

which means that $\frac{1}{h}|S_1(h)| \to 0$ when h tends to zero.

The estimates for $|S_2(h)|$ and $|S_3(h)|$ involve only the localization of the wavelets. The regularity plays no further role since the wavelets whose support contain t_0 are zero in a neighborhood of $t_0 + h$ and conversely.

We take absolute values inside the sum and use (9.8) while distinguishing those values of k for which $\psi_{j,k}(t_0) \neq 0$ and those for which $\psi_{j,k}(t_0 + h) \neq 0$. In the second case, $|k2^{-j} - t_0|$ is the order of magnitude of h.

The contribution $|S_2(h)|$ is thus bounded by a constant times

$$(9.13) \qquad \sum_{j_0 < j \leq j_1} 2^{-j} \eta(2^{-j}) + (j_1 - j_0)h\eta(h) + h \sum_{j_0 < j \leq j_1} \eta(2^{-j}).$$

We have assumed that $\int_0^1 \eta(t)\frac{dt}{t} < \infty$ and that $\eta(t)$ is an increasing function. From this it follows that $\sum_0^\infty \eta(2^{-j}) < \infty$. Furthermore, $\eta(2^{-j})$ decreases to 0, from which we conclude that $\lim_{j\to+\infty} j\eta(2^{-j}) = 0$ and, hence, that $(\log\frac{1}{h})\eta(h) \to 0$. But $j_1 - j_0$ is of the order $\log\frac{1}{h}$ with the result that $\lim_{h\to 0}\frac{1}{h}S_2(h) = 0$.

To bound $|S_3(h)|$, we use only the hypothesis on the global regularity of f as expressed in terms of the wavelet coefficients (9.3). We again take absolute values inside the sum. By considering, as in the second case, the limitations imposed by the compact support of ψ, we deduce that $|S_3(h)| \leq Ch^2$, which implies that $\lim_{h\to 0}\frac{1}{h}S_3(h) = 0$.

This completes the proof of Theorem 9.1; however, this is not the result that will be used in the next section. We will need a "corollary of the proof" of Theorem 9.1. The statement of the corollary itself follows.

COROLLARY. *Let $\theta(x,y)$, $x \in \mathbb{R}$, $0 < y \leq 1$, be a measurable function of (x,y) satisfying, in addition, the following two conditions:*

$$(9.14)\qquad y|\theta(x,y)| \leq Cy^\alpha \quad \text{for all} \quad x \in \mathbb{R} \quad \text{and all} \quad y \in (0,1]$$

for some constant C and an exponent $\alpha > 0$ and

$$(9.15)\qquad y|\theta(x,y)| \leq y\eta(y) + |x - t_0|\eta(|x - t_0|),$$

where the function η satisfies the hypotheses of Theorem 9.1, in particular (9.9).

Suppose that $g(t)$ is a function in the class C^2 with compact support. Define

$$(9.16)\qquad f(t) = \int_0^1 \int_{-\infty}^\infty \theta(x,y)g\left(\frac{t-x}{y}\right) dx\frac{dy}{y}.$$

Then $f(t)$ is differentiable at t_0, and $f'(t_0)$ can be computed by differentiating under the integral sign:

$$(9.17)\qquad f'(t_0) = \int_0^1 \int_{-\infty}^\infty \theta(x,y)g'\left(\frac{t_0-x}{y}\right) dx\frac{dy}{y^2},$$

and the integral (9.17) is absolutely convergent.

The proof of the corollary is parallel to that of Theorem 9.1. Observe that if we examine the absolute convergence of the integral (9.17) without considering (9.14), then we come directly to the Dini condition (9.9).

9.4.   Study of the Riemann function.

According to historians, Karl Weierstrass mentioned the function $W(t) = \sum_1^\infty \frac{1}{n^2}\sin(\pi n^2 t)$ in a talk to the Academy of Sciences in Berlin on 18 July 1872 and indicated that Riemann had introduced this function to warn mathematicians that a continuous function need not have a derivative anywhere [5]. This function has come to be known as "Riemann's function," although there seems to be no written evidence that connects either Riemann or his students with this function. (See [4] for a fascinating discussion of the mystery surrounding the origin of $W(t)$.)

Not being able to prove that $W(t)$ was nowhere differentiable, Weierstrass considered the more lacunary series $\sigma(t) = \sum_0^\infty b^n \cos(a^n t)$, $0 < b < 1$, and he showed that if a is an odd integer and if ab is sufficiently large, then $\sigma(t)$ is nowhere differentiable. We have seen that the result is true for $a = 2$ and $b = 1/2$; in fact, $ab \geq 1$ is sufficient.

In 1916, G. H. Hardy proved that the Riemann function $W(t)$ is not differentiable at t_0 in the following three cases:

(9.18) t_0 is irrational;

(9.19) $t_0 = p/q$ with $p \equiv 0 \pmod 2$ and $q \equiv 1 \pmod 4$;

(9.20) $t_0 = p/q$ with $p \equiv 1 \pmod 2$ and $q \equiv 2 \pmod 4$.

Serge Lang routinely suggested this "conjecture of Riemann" to his students and, to the general surprise of the mathematics world, Joseph L. Gerver, one of Lang's undergraduate students, resolved the problem by proving the following unexpected result: If $t_0 = p/q$ where p and q are odd, then $W(t)$ is differentiable at t_0 and $W'(t_0) = -1/2$. He then showed that $W(t)$ is differentiable at no other points, and the problem of the differentiability of the Riemann function was completely settled.

Following Holschneider and Tchamitchian [8], we restudy the function $W(t)$ by associating with it the corresponding "analytic signal" defined by

(9.21) $$F(t) = \sum_1^\infty \frac{1}{n^2} e^{i\pi n^2 t}.$$

$F(t)$ can be extended to the upper-half plane $z = x + iy$, $y > 0$, as the bounded, holomorphic function

(9.22) $$F(z) = \sum_1^\infty \frac{1}{n^2} e^{i\pi n^2 z}.$$

The most natural wavelet analysis for such holomorphic functions is the analysis advocated by Lusin (§2.5). The wavelet transform of $F(z)$ is, up to a normalization, $F'(z) = i\pi \sum_1^\infty \exp(i\pi n^2 z) = \frac{i\pi}{2}(\theta(z) - 1)$, where $\theta(z)$ is the *Jacobi function* defined by

(9.23) $$\theta(z) = \sum_{-\infty}^\infty \exp(i\pi n^2 z), \qquad z = x + iy, \qquad y > 0.$$

It is easy to study the behavior of the function $\theta(z)$ near the real axis by using the functional equation satisfied by $\theta(z)$. And everything would go well in the best of all possible worlds if the analyzing wavelet, $\psi(t) = (t + i)^{-2}$, satisfied the minimal condition allowing the corollary of Theorem 9.1 to be used, namely, that $\int_{-\infty}^\infty (1 + |t|)|\psi(t)|dt < \infty$. But this integral diverges.

Holschneider had the remarkable idea to start with the result of the analysis obtained with the "bad wavelet," and then to do the synthesis with the "good wavelet" $g(t)$. Thus he considered a function $g(t)$, in the class C^2 and with compact support, such that

(9.24) $$\int_0^\infty \hat{g}(u)e^{-u}du = 1,$$

where $\hat{g}$ denotes the Fourier transform of g. We then have, for all $\lambda > 0$,

(9.25) $$\int_0^\infty \int_{-\infty}^\infty g\left(\frac{t-x}{y}\right) e^{i\lambda(x+iy)}dx\frac{dy}{y} = \frac{1}{\lambda}e^{i\lambda t}$$

and

(9.26) $$\int_0^\infty \int_{-\infty}^\infty g\left(\frac{t-x}{y}\right)(\theta(x+iy)-1)dx\frac{dy}{y} = \frac{2}{\pi}\sum_1^\infty \frac{e^{i\pi n^2 t}}{n^2}.$$

We can further simplify the identity (9.26) by requiring that the integral of $g(t)$ be zero, which is to say that it is a Grossmann–Morlet wavelet. Then (9.26) becomes

(9.27) $$\int_0^\infty \int_{-\infty}^\infty g\left(\frac{t-x}{y}\right)\theta(x+iy)dx\frac{dy}{y} = \frac{2}{\pi}\sum_1^\infty \frac{e^{i\pi n^2 t}}{n^2}.$$

Let $t_0 = \frac{2p+1}{2q+1}$. To demonstrate that the second member of (9.27) is differentiable at t_0, we use the corollary of Theorem 9.1.

First of all, the contribution from $y \geq 1$ in (9.27) can be neglected. In fact, $|\theta(z) - 1| \leq Ce^{-y}$ when $y \geq 1$ and $z = x + iy$ so that the integral in y from 1 to infinity yields a function in the class C^2 or, more generally, having the same regularity as $g(t)$.

To establish the differentiability at t_0, observe that

(9.28) $\qquad |\theta(x+iy)| \leq Cy^{-1/2}$ $\quad$ for all real $\quad x \quad$ and $\quad 0 < y \leq 1$

and that

(9.29) $\qquad y|\theta(t_0 + z)| \leq C|z|^{3/2}$ $\quad$ if $\quad z = x + iy, \qquad 0 < y \leq 1.$

We can then apply the corollary of Theorem 9.1 and prove Gerver's theorem.

9.5. Conclusions.

In view of the example of the Riemann function, one would be tempted to conclude that wavelet analysis is better than Fourier analysis for studying fractal structures. The Riemann function is given explicitly in terms of a Fourier series and, yet, this exact spectral information yields no direct access to the pointwise regularity of the function. This conclusion, however, is incorrect for several reasons.

First of all, the simplest and most precise method for studying the Riemann function has been found by Itatsu [9], and *it is not a time-scale method*. On the contrary, Itatsu's method is based on the judicious use of the Poisson summation formula. It is thus a "time-frequency" method, and it gives *more precise* results than those obtained by wavelet analysis.

In the second place, we know today that wavelet analysis is part of a much larger discipline, namely, 2-microlocal analysis ([2], [10]). 2-microlocal analysis saw the light of day at about the same time as wavelet analysis. The dictionary allowing us to go from one type of analysis to the other has been worked out by Jaffard [10].

Should one use wavelet analysis or 2-microlocal analysis? We refer the reader to [11], where this problem is carefully studied. My (subjective) point of view is that 2-microlocal analysis is a more flexible and more precise instrument but that its sophistication can discourage some scientists. Wavelet analysis offers an apparent simplicity, which is one of the explanations of its success.

Bibliography

[1] A. ARNÉODO, E. BACRY, AND J. F. MUZY, *Wavelets and multifractal formalism for singular signals: Applications to turbulence data*, preprint, Centre de Recherche Paul Pascal, Pessac, France, 1991.

[2] J. M. BONY, *Second microlocalization and propagation of singularities for semi-linear hyperbolic equations*, Taniguchi Symp. HERT. Katata, (1984), pp. 11–49.

[3] ———, *Interaction des singularités pour l'équation de Klein-Gordon non linéaire*, Séminaire Goulaouic 1983-1984, Ecole Polytechnique, Centre de Mathématique, Palaiseau, France.

[4] P. L. BUTZER AND E. L. STARK, *"Riemann's example" of a continuous nondifferentiable function in the light of two letters (1865) of Christoffel to Prym*, Bull. Soc. Math. Belg., 38 (1986), pp. 45–73.

[5] J. J. DUISTERMAAT, *Selfsimilarity of 'Riemann's nondifferentiable function,'* Nieuw Arch. Wisk., 9 (1991), pp. 303–337.

[6] J. GERVER, *The differentiability of the Riemann function at certain rational multiples of π*, Amer. J. Math., 92 (1970), pp. 33–55.

[7] ———, *More on the differentiability of the Riemann function*, Amer. J. Math., 93 (1970), pp. 33–41.

[8] M. HOLSCHNEIDER AND PH. TCHAMITCHIAN, *Pointwise analysis of Riemann's "non differentiable" function*, Inventiones Mathematicae, 105 (1991), pp. 157–176.

[9] S. ITATSU, *The differentiability of the Riemann function*, Proc. Japan Acad., Ser. A, Math. Sci. 57 (1981), pp. 492–495.

[10] S. JAFFARD, *Pointwise smoothness, two microlocalization and wavelet coefficients*, Publicacions Mathemàtiques (Publicacions de la Universitat Autònoma de Barcelona), 35 (1991), pp. 155–168.

[11] Y. MEYER, *L'analyse par ondelettes d'un objet multifractal: la function $\sum_1^\infty \frac{1}{n^2} \sin n^2 t$ de Riemann*, Colloquium Mathématique de l'Université de Rennes, Rennes, France, November 1991.

[12] H. QUEFFELEC, *Dérivabilité de certaines sommes de séries de Fourier lacunaire*, Thesis, University of Orsay, Orsay, France, 1971.

CHAPTER 10

Wavelets and Turbulence

10.1. Introduction.

The study of profound problems is often influenced by the available instruments and techniques. The example of Galileo's lens comes immediately to mind. We propose, by taking up an argument due to Marie Farge, to show that it is as natural to study turbulence using wavelets as it is to explore the night sky with a telescope.

This is not to imply, by any means, that all the problems in turbulence have been resolved thanks to wavelets. We merely see things a little more clearly. The objective of this chapter is to describe the new point of view that wavelets have brought to the study of fully developed turbulence.

The following lines are, in reality, only "lecture notes," and we encourage those who wish to know more about this subject to look at the original articles cited at the end of the chapter.

10.2. The statistical theory of turbulence and Fourier analysis.

The statistical theory of turbulence was introduced more or less simultaneously by Kolmogorov (1941), Obukhov (1941), Onsager (1945), Heisenberg (1948), and Von Weizsäcker (1948). This work involved applying the statistical tools used for studying stationary process to understand the partition of energy, at different scales, in the solutions of the Navier–Stokes equation.

According to Leray [6], this statistical point of view if justified by the loss of stability and uniqueness of the solutions for very large Reynolds numbers and for large values of time. One then speaks of fully developed turbulence.

The intermediate scales (the inertial zone) lie between the smallest scales (where, through viscosity, the dynamic energy is dissipated in heat) and the largest scales (where exterior forces supply the energy). In this inertial zone, the theory of Kolmogorov stipulates that energy is neither produced nor dissipated but only transferred, without dissipation, from one scale to another and according to a constant rate ε.

Another hypothesis is that turbulence is statistically homogeneous (invariant under translation), isotropic (invariant under rotation), and self-similar. The velocity components are treated as random variables, in the probabilistic sense, and the statistical description is derived from the corresponding correlation functions. The mathematical tool adapted to this statistical approach is the Fourier transform, and, by associating scale and frequency in the usual way, Kolmogorov and Obukhov arrived at $\varepsilon^{2/3}|k|^{-5/3}$ for the average spectral distribution of energy, where k is the vector-variable of the Fourier transform, which was taken over the space variables x_1, x_2, and x_3.

Various wind-tunnel experiments [Batchelor and Townsend (1940); and Anselmet, Gagne, Hopfinger, and Antonio (1984)] have shown that the energy associated with the small scales of a turbulent flow is not distributed uniformly in space. This observation, that the support of the transfer of energy is spatially intermittent, has led several authors to hypothesize that this support is fractal [Mandelbrot (1975); and Frisch, Sulem, and Nelkin (1978)] or multifractal [Parisi and Frisch (1985)].

Use of the Fourier transform does not elucidate the multifractal structure of fully developed turbulence, and this observation leads us to an important application of the wavelet transform.

10.3.   Verification of the hypothesis of Parisi and Frisch.

The conjecture of Parisi and Frisch [9] is based on experimental data supplied by Anselmet, Gagne, Hopfinger, and Antonio. The computation that led to their conjecture involved evaluating, as a function of the displacement $|\Delta x|$, the average value of the pth power of the change in the vector velocity in a turbulent fluid, that is the average value of $|v(x + \Delta x, t) - v(x, t)|^p$. The surprising result from these measurements was that they obtained a power law in terms of $|\Delta x|^{\zeta(p)}$, where the exponent $\zeta(p)$ does not depend linearly on p. The interpretation given by Parisi and Frisch is that turbulent flow develops multifractal singularities when the Reynolds number becomes very large.

The relation between the multifractal structure and the power law is given by the following heuristic reasoning. To speak of a multifractal structure means that, for each $h > 0$, there is a set of singular points with Hausdorff dimension $D(h)$ on which the increase in velocity acts like $|\Delta x|^h$. The contribution of these "singularities of exponent h" to the average value of $|v(x + \Delta x, t) - v(x, t)|^p$ is of the order of magnitude of the product $|\Delta x|^{ph}|\Delta x|^{3-D(h)}$; the second factor is the probability that a ball of radius $|\Delta x|$ intersects a fractal set of dimension $D(h)$.

When $|\Delta x|$ tends to 0, the dominant term is the one with the smallest possible exponent, which leads to

$$\zeta(p) = \inf_{h>0} \{ph + 3 - D(h)\}.$$

The exponent $\zeta(p)$ is thus given by the Legendre transform of function $D(h)$, which is, we recall, the Hausdorff dimension of the set of exceptional point x, where $|v(x + \Delta x, t) - v(x, t)|$ is the order of magnitude of $|\Delta x|^h$. The nonlinear dependence of $\zeta(p)$ on p thus indicates that the very abrupt variations in velocity correspond to a multifractal structure.

The wavelet transformation is the ideal tool for analyzing multifractal structures, and Frisch has sought to verify his conjecture by plunging into the very heart of the turbulent signal and traveling across the scales to calculate the fractal exponents.

The turbulent signal, which is the object of the analysis, is, in fact, a function of time obtained by hot-wire measurements. The turbulent flow comes from the Modane wind tunnel, and the hot-wire technique provides the measurement at a given point of the velocity as a function of time. Taylor's hypothesis implies that, for the particular conditions created in this tunnel, the "segment in time" is equivalent to a "segment in space" (along the axis of the tunnel). Finally, the turbulent signal is analyzed with the wavelet transform and, thanks to a two-dimensional color visualization, one displays a multifractal structure.

This phenomenological approach has been criticized by Everson, Sirovich, and Sreenivasan [3]. These investigators have been able to show that wavelet analysis of Brownian motion produces very similar two-dimensional visualizations.

It then became a matter of urgency to move from the qualitative to the quantitative and to extract the fractal exponents h and the corresponding Hausdorff dimensions $D(h)$ from the "Gagne signal." This research is being actively pursued by Alain Arnéodo and his collaborators with the objective to confirm the hypothesis of Frisch and Parisi.

10.4. Farge's experiments.

Following Farge, we quote John von Neumann [10]:

The phenomenon of turbulence was discovered physically and is still largely unexplored by mathematical techniques. At the same time, it is noteworthy that the physical experimentation which leads to these and similar discoveries is a quite peculiar form of experimentation; it is very different from what is characteristic in other parts of physics. Indeed, to a great extent, experimentation in fluid dynamics is carried out under conditions where the underlying physical principles are not in doubt, where the quantities to be observed are completely determined by known equations. The purpose of the experiment is not to verify a proposed theory but to replace a computation from an unquestioned theory by direct measurements. Thus wind tunnels are, for example, used at present, at least in large part, as computing devices of the so-called analogy type (or, to use a less widely used, but more suggestive, expression proposed by Wiener and Caldwell: of the measurement type) to integrate the nonlinear partial differential equations of fluid dynamics.

Thus it was to a considerable extent a somewhat recondite form of computation which provided, and is still providing, the decisive mathematical ideas in the field of fluid dynamics. It is an analogy (i.e., measurement) method, to be sure. It seems clear, however, that digital (in the Wiener–Caldwell terminology: counting) devices have more flexibility and more accuracy, and could be made much faster under present conditions. We believe, therefore, that it is now time to concentrate on effecting the transition to such devices, and that this will increase the power of the approach in question to an unprecedented extent.

One of the chief architects of this experimentation has, without doubt, been Norman Zabusky, who, after having discovered solitons (in collaboration with

Kruskal), demonstrated *the existence of coherent structures within turbulent flows in two as well as three dimensions.* Zabusky comments on his discovery [11]:

In the last decade we have experienced a conceptual shift in our view of turbulence. For flows with strong velocity shear ... or other organizing characteristics, many now feel that the spectral or wavenumber-space description has inhibited fundamental progress. The next "El Dorado" lies in the mathematical understanding of coherent structures in weakly dissipative fluids: the formation, evolution and interaction of metastable vortex-like solutions of nonlinear partial differential equations...

Farge explains to us how and why she was led to use wavelet analysis in her study of numerical simulations of two-dimensional fully developed turbulence [4]:

The use of the wavelet transform for the study of turbulence owes absolutely nothing to chance or fashion but comes from a necessity stemming from the current development of our ideas about turbulence. If, under the influence of the statistical approach, we had lost the need to study things in physical space, the advent of supercomputers and the associated means of visualization have revealed a zoology specific to turbulent flows, namely, the existence of coherent structures and their elementary interactions, none of which are accounted for by the statistical theory...

What Farge asks of wavelet analysis (or of any other form of "time-frequency" analysis) is to decouple the dynamics of the coherent structures from the residual flow. The residual flow would play only a passive role in an action whose protagonists would be the coherent structures; these "protagonists" clash or join forces according to their "sign"...

The difficulty arises because the Navier–Stokes equations are nonlinear; hence, the interactions between the coherent structures and the residual flow cannot be eliminated. In other words, the coherent structures differ from solitons in that they are not particular solutions of the Navier–Stokes equations.

Farge, after having tried several methods to extract the coherent structures from the residual flow, decided to use Victor Wickerhauser's algorithm, which provides a decomposition in a basis adapted to the signal. It is quite remarkable that the Wickerhauser algorithm extracted the coherent structures by giving them priority over the residual flow. These unexpected results are in the process of being published.

10.5. Numerical approaches to turbulence.

It is not unreasonable to believe that the use of new methods in numerical analysis will considerably reduce the time needed to compute solutions of the Navier–Stokes equations.

Gregory Beylkin is developing a systematic program in which wavelet analysis replaces the more traditional methods of numerical analysis—finite elements, finite differences, and spectral methods. He intends to apply these methods to the solution of the Navier–Stokes equation.

Farge poses the following questions in [4]:

1. Is it possible to "project," in the sense of Galerkin's method, the Navier-Stokes equations onto bases that are adapted to the structure of these equations and that lead to efficient numerical computation?

2. More generally, can one effectively describe the solutions with a small number of parameters, as is the objective of the theory of inertial varieties of Foïas and Temam? These inertial varieties contain the "strange attractors," which constitute the asymptotic solutions.

3. Can the best basis method of Wickerhauser furnish an effective parameterization for these inertial varieties?

There are many questions that have yet to receive satisfactory answers, but these issues are stimulating research projects that will perhaps be decisive.

Bibliography

[1] F. ARGOUL, A. ARNÉODO, G. GRASSEAU, Y. GAGNE, E. F. HOPFINGER, AND U. FRISCH, *Wavelet analysis of turbulence reveals the multifractal nature of the Richardson cascade*, Nature 338 (1989), pp. 51–53.

[2] A. ARNÉODO, G. GRASSEAU, AND M. HOLSCHNEIDER, *Wavelet transform of multifractals*, Phys. Rev., 61 (1988), pp. 2281–2284.

[3] R. EVERSON, L. SIROVICH, AND K. R. SREENIVASAN, *Wavelet analysis on the turbulent jet*, Phys. Lett. A, 145 (1990), pp. 314–324.

[4] M. FARGE, *Transformée en ondelettes continue et application à la turbulence*, Société Mathématique Française, Paris, France, May 5, 1990.

[5] A. M. KOLMOGOROV, *A refinement of previous hypotheses concerning the local structure of turbulence in viscous incompressible fluid at a high Reynolds number*, J. Fluid Mech., 13 (1961), pp. 1, 82–85.

[6] J. LERAY, *Etudes de diverses équations intégrales non-linéaires et de quelques problèmes que pose l'hydrodynamique*, J. Math. Pures Appl., (1933), pp. 1–82.

[7] B. MANDELBROT, *Intermittent turbulence in self-similar cascades: Divergence of high moments and dimension of carrier*, J. Fluid Mech., 62 (1975), pp. 331–358.

[8] J. F. MUZY, E. BACRY, AND A. ARNÉODO, *Wavelets and multifractal formalism for singular signals: Application to turbulence data*, preprint, Centre de Recherche Paul Pascal, Pessac, France.

[9] G. PARISI AND U. FRISCH, *Turbulence and predictability in geophysical fluid dynamics and climate dynamics*, M. Ghil, R. Benzi, and G. Parisi, eds., North-Holland, Amsterdam, 1985, pp. 71–88.

[10] J. VON NEUMANN, *Complete works*, 1949. [Translator's note: The quotation is from *On the Principles of Large Scale Computing Machines* by Herman H. Goldstine and John von Neumann. This paper was never published. It contains material given by von Neumann in a number of lectures, in particular one at a meeting on May 15, 1946, of the Mathematical Computing Advisory Board, Office of Research and Inventions, Navy Department, which in 1947 became the Office of Naval Research.]

[11] N. ZABUSKY, *Computational synergetics*, Physics Today, July (1984), pp. 2–11.

CHAPTER 11

Wavelets and the Study of Distant Galaxies

11.1. Introduction.

Wavelets are being used by Albert Bijaoui and his collaborators to clarify the hierarchical organization of distant galaxies and, possibly from this, to make deductions about the formation of these galaxies.

The following lines are again "lecture notes," this time from original articles by Bijaoui and others, and we encourage the interested reader to consult these articles. Several are listed in the bibliography.

11.2. The new telescopes.

The difficulties encountered today in the analysis of the distribution of the galaxies result, paradoxically, from technological progress in the construction of telescopes.

On the one hand, the quality of telescopes has been considerably improved during the last 50 years. In 1950, astronomers could capture ten million galaxies. Today, 100 million galaxies can be examined... There is now such a quantity of information that it is impossible to continue using traditional observations based on the astronomer's eye and judgment.

On the other hand, the very nature of the images coming from these telescopes has undergone a revolution. Now, CCDs (Charge Coupled Devices) replace plates of silver salts, and "chemical" photography is already a thing of the past. The future will involve electronic reception of photons that come to us from the edge of the Universe.

CCDs do not provide images in the traditional sense but, rather, data that are suitable for various methods of processing. Tomorrow's telescopes will be computers that can be accessed from a distance. These computers will, with proper software, automatically execute certain image-processing algorithms. The optimal utilization of the marvelous progress offered by these CCDs necessitates implementing effective algorithms for processing the received images, and this is where wavelets enter.

11.3.   The hierarchical organization of the galaxies and the creation of the Universe.

This considerable technological progress has allowed us to study the three-dimensional organization of the galaxies. The results that have been obtained challenge the imagination. Far from being arbitrarily distributed like points tossed at random, the galaxies are organized according to extremely complex geometric configurations, and these configurations contain information. In some cases, galaxies are distributed in filaments; in other cases, they are spread over huge surfaces that surround only void.

Galaxies aggregate into clusters, which are themselves organized into super-clusters, although a galaxy can belong to a supercluster without being part of a cluster.

One of the goals of contemporary astronomy is to analyze, in detail, the geometry of these hierarchical galactic organizations. This analysis is particularly important in the case of the most distant galaxies. Indeed, the images that we receive from these galaxies date from the beginning of the creation of the Universe. It is possible to imagine that the fascinating geometry of these hierarchical organizations of distant galaxies preserves traces of the process of "fragmentation" of the smooth, homogeneous primitive matter that made up the Universe.

11.4.   The multifractal approach to the Universe.

A fractal description of the distribution of matter in the Universe has been proposed by Mandelbrot (1975, 1982) and, more or less explicitly, by many other scientists. But the nature of the distribution of the galaxies precludes homogeneous fractal descriptions, whereas a multifractal approach corresponds with reality [7].

Although it gives no indication of the physical processes of creation, this multifractal approach would, however, be very useful if it led to phenomenological predictions about certain aspects of the galaxies' distribution—the frequency of voids, etc... But the scientists who were using these "multifractal" approaches were oriented toward calculating global information and parameters, and they were neglecting the study of the local fractal structure of the galaxies.

11.5.   The advent of wavelets.

We encounter the same situation here as the one that arose in the study of turbulence. In that case, the passage from global fractal properties to local fractal properties required the use of wavelets. The hope placed in the analysis of astronomic images using wavelets is, without doubt, reasonable if one considers the arguments presented by Marr. Wavelet analysis is used to delimit the boundaries of objects and, by so doing, to arrive at their three-dimensional organization. But it is precisely this three-dimensional organization that is sought by the astrophysicists.

On the other hand, at a more imaginative level, it is tempting to compare the flow of the Universe over time to that of a fluid and to continue this metaphor by comparing the galaxies to the active zones of fully developed turbulence. Wavelet analysis of astronomic images is thus as natural as the wavelet analysis of turbulence.

We hope, by these few lines, to have stimulated the readers' curiosity and, in this case, we encourage him or her to go to the original scientific articles by Bijaoui, Slezak, and others.

Bibliography

[1] PH. BENDJOYA, E. SLEZAK, AND CL. FROESCHLÉ, *The wavelet transform, a new tool for astroid family determination*, Astronom. and Astrophys., to appear.

[2] A. BIJAOUI, *Algorithmes de la transformation en ondelettes*, Application à l'image astronomique, Proc. du Cours CEA/EDF/INRIA, 1991, pp. 1–26.

[3] ——, *Le ciel lointain est-il un mirage?* Ciel et Espace, numéro spécial "Du big-bang à nos jours," June-July-August 1991.

[4] A. BIJAOUI AND M. GIUDICELLI, *Optical image addition with the wavelet transform*, Experiment. Astron., 1 (1991), pp. 347–363.

[5] A. BIJAOUI, E. SLEZAK, AND G. MARS, *The wavelet transform: A new way to describe the Universe*, Workshop on the Distribution of Matter in the Universe, Meudon, France, March 1991.

[6] E. ESCALERA, E. SLEZAK, AND A. MAZURE, *New evidence for subclustering in the Coma clustering using the wavelet analysis*, Astronom. and Astrophys., 269 (1992), pp. 379–384.

[7] B. J. T. JONES, V. J. MARTINEZ, E. SAAR, J. EINASTO, *Multifractal description of the large-scale structure of the universe*, Astrophys. J., 332 (1988), pp. 1–5.

[8] E. SLEZAK, A. BIJAOUI, AND G. MARS, *Identification of structures from galaxy counts: Use of the wavelet transform*, Astronom. and Astrophys., 227 (1990), pp. 301–316.

Index

adaptive filtering. *See* wavelet packets
adaptive segmentation. *See* Malvar wavelets
Adelson. *See also* pyramid algorithms
    image processing, 34, 45
analytic signal, 23. *See also* Wigner–Ville
    transform
    instantaneous frequency, 70–71
Arnéodo
    Frisch and Parisi hypothesis, 17, 121
atomic decompositions
    Calderón's identity, 24
    Hardy spaces, 23–24, 26
    $L^p[0,1]$, 24
    wavelet analysis, 31
    Wigner–Ville transform, 73
atoms, 4–5. *See also* time-frequency atoms

Balian
    time-frequency representation, 63
Balian–Low theorem, 35, 76–77
Battle
    renormalization, 31
Benassi
    Gaussian–Markov fields, 19
best basis. *See also* Malvar wavelets
    approach to turbulence, 123
    Malvar wavelets, 83–84
    wavelet packets, dual to Malvar wavelets,
        99
Beylkin
    wavelets and numerical analysis, 122
bi-orthogonal
    pyramids. *See* pyramid algorithms
    wavelets, 59–61
Bijaoui
    wavelet analysis in astronomy, 111, 125
Boulez
    Mozart's *Magic Flute*, 65
Brownian motion, 18–19
Burt and Adelson's algorithm. *See* pyramid
    algorithms.

Calderón, 13
    identity
        decomposition of the identity, 25

Grossmann–Morlet theory, 28
    rediscovered by Grossmann and
        Morlet, 13
Charge Coupled Devices
    in astronomy, 125
chirp signals, 68–69, 72
Ciesielski, 22
coding. *See also* signal and image processing
    Burt and Adelson's algorithm, 52, 55
    bi-orthogonal wavelets, 61
    linear prediction coding, 33
    Mallat's "herringbone" algorithm, 39
    Mallat's image coding algorithm, 106
    orthogonal pyramids, 57
    subband coding, 3, 34–36
    transform coding, 3, 33–34
    trends and fluctuations, 55, 57–58
    zero-crossings, 3, 13, 101
Cohen
    bi-orthogonal wavelets, 59–60
    convergence of Mallat's algorithm, 58
coherent states, 25
Coifman. *See also* Weiss
    Hardy spaces, 26–27
    Lusin's theory, 24
    time-frequency bases, 77
Coifman–Weiss, 26–27
compression. *See also* signal and image
        processing
    asymptotic limits of coding schemes, 34
    optimal with Malvar basis, 83
    pyramid algorithms, 51
Croissier
    quadrature mirror filters, 30, 33
cubic splines, 48, 107–108

Daubechies
    bi-orthogonal wavelets, with Cohen and
        Feauveau, 59–60
    extended Haar's work, 6, 30
    time-frequency wavelet, with Jaffard and
        Journé, 77
    wavelets, 8, 10, 29–30